STATISTICS BY EXAMPLE
Weighing Chances

STATISTICS BY EXAMPLE
Weighing Chances

Prepared and edited by the
Joint Committee on the Curriculum in Statistics and Probability
of the American Statistical Association
and the National Council of Teachers of Mathematics

Frederick Mosteller, *Chairman*
Harvard University

Richard S. Pieters
Phillips Academy, Andover

William H. Kruskal
The University of Chicago

Gerald R. Rising
State University of New York at Buffalo

Richard F. Link
Artronic Information Systems, Inc.

with the assistance of

Roger Carlson
University of Missouri at Kansas City

Martha Zelinka
Weston High School, Weston, Massachusetts

ADDISON-WESLEY PUBLISHING COMPANY
Reading, Massachusetts · Menlo Park, California · London · Don Mills, Ontario

This book is in the
Addison-Wesley Innovative Series

ISBN 0-201-04875-2
EFGHIJK—EB—83210

Preface

Examples of statistical activity, problems, and conclusions are abundant. They appear in newspapers and magazines, in textbooks and television, and in almost every aspect of life.

Nonetheless, clear, interesting, and elementary written descriptions of such examples are hard for students and teachers to find. The reasons are various. Many examples are large in scale and require facilities for extensive handling of data. Sometimes the statistical aspects of a real-life problem appear in the middle of its solution, and when removed from this context may be hard to appreciate. The statistical part may even be invisible unless detected by an expert. Again, our tendency in mathematics teaching is to boil each problem down to an abstract skeletal minimum. Real-life statistical problems usually have extramathematical features as well as mathematical ones, and the abstraction is not ordinarily already done for the user.

In preparing the present material, the Joint Committee on the Curriculum in Statistics and Probability of the American Statistical Association and the National Council of Teachers of Mathematics thought it wise to move strongly in the direction of real-life problems and to admit the need for more explanatory text than we find in the usual mathematics book. We feel that there is a need to explain real-life problems because the student is unlikely to have had experience with any kind of statistics beyond taking averages, and the teacher with any other than the theoretical side of statistics, if that.

This series of pamphlets, *Statistics by Example*, has four parts, each intended to stand alone, but cross-referencing is given. The subtitles and reading needs of the parts are

Exploring Data	Arithmetic, rates, percentages
Weighing Chances	Notion of probability, elementary algebra
Detecting Patterns	Elementary probability, intermediate algebra
Finding Models	Elementary probability, intermediate algebra

The tables of contents and the introduction to each part give much more detail than the following brief sketches.

Exploring Data shows how to organize data tabularly and graphically to get messages and exhibit them, and it introduces elementary probability in circumstances where simple counting gets one off to a good start.

Weighing Chances develops probability methods through random numbers, simulations, and simple probability models, and presents serious analyses of complex data using informally the idea of scatter and residuals.

Detecting Patterns presents several standard statistical devices—the normal distribution, the chi-square test, and regression methods.

Finding Models encourages the student to develop models as structures for data, so that departures from the models can be recognized and new structures built.

In all four parts the keynote is real problems with real data. A few problems are treated in two separate parts, at different levels, and usually without heavy overlap.

In making cross-references to treatments in other parts, we refer to:

SBE, Subtitle, Set number.

Here *SBE* is an abbreviation of *Statistics by Example.*

We present here a body of material very different from the usual. We have, for example, not hesitated to lean occasionally upon sports in this discussion because it helps some students to learn about a statistical method in a sports context that he knows about. Then he has a chance to raise the honest "but" 's that are a part of any serious statistical analysis. On the other hand, we rarely treat well-shuffled cards or balls drawn from urns, though these are a useful stock in trade of the beginning probability student. Instead, we have tried to work with less sterilized problems.

These volumes are not mere collections of problems and examples. Rather, each represents a series of mini-learning experiences or case histories and includes exercises and projects relative to the examples.

Our material, including the exercises and projects, should help students and teachers learn to analyze statistical problems. The projects represent larger ventures that an individual or a class can carry out. By so doing, a student will learn about the practical as well as the analytic problems associated with statistical work.

Some of the sets of examples may fit well in courses other than mathematical ones—social science, biology, physics, English, economics, civics, and history.

Those who enjoy self-study may find especially congenial the approach of learning by example.

Beyond the materials we offer here, the Joint Committee has also prepared a book of descriptions of a variety of important applications entitled *Statistics: A Guide to the Unknown* (Judith Tanur *et al.*, editors, Holden-Day Publishing Co., 500 Sansome St., San Francisco, Calif., 1972). Consequently, in addition to being able to begin learning how to do statistics through examples, the student and the teacher can supplement their work by reading about applications to nearly any field of man's work and life.

How was the present book, *Statistics by Example*, prepared? The book is composed of sets of examples based upon real problems, written up by professional statisticians and experienced teachers. After they were drafted, they were sent to teachers who reviewed them for their understandability and often asked their students for furthur comments. On the basis of the criticism received, and additional critiques from members of the committee, the example sets were revised, usually more than once. The names of the teachers who formed this task force are listed below.

We owe a great debt to the teachers who provided us with so much constructive criticism. And we owe much to the authors who patiently revised their work to make it more comprehensible. The sole reward of both groups is that they have provided this new material for teaching statistics to the community.

The Committee members have served as editors for the material, and we have had the editorial assistance of Roger Carlson and Martha Zelinka.

Robert Berk, H. T. David, Janet Elashoff, Chien Pai Han, David Hoaglin, Robert Kleyle, and Perry Gluckman have helped us in a variety of ways.

We have been aided secretarially by Muriel Ferguson, Holly Grano, and Janet Quint, who also helped with the editing, as did Cleo Youtz. Holly Grano brought her expert skill to the final manuscript shown here.

Support for the Committee's work was generously provided by the Sloan Foundation.

September 1, 1972   Frederick Mosteller, Chairman
                                                  William H. Kruskal
                                                  Richard F. Link
                                                  Richard S. Pieters
                                                  Gerald R. Rising

Task Force Members

James Arnold
Thomas C. Armstrong
E. W. Averill
Crayton Bedford
Max S. Bell
James Bierden
Shirley Brady
Richard G. Brown
William G. Chinn
Jack B. Eleyet
George A. Fargo
D. Fishman
Don Fraser
Abraham M. Glass
Alice Golar
Nicholas Grant
George Grossman
Robert L. Heiny
G. Albert Higgins, Jr.
Gertrude Hillman
Donald K. Hotchkiss

John W. Hulse
Warren O. Hulser
S. Ishikawa
Ralph H. Klitz, Jr.
George C. Laumann
M. Albert Linton, Jr.
John Malak
Arthur J. Mastera
Margaret Matchett
William K. McNabb
James V. Mulloy
Christine Murphy
Fernand J. Prevost
Raymond E. Roth
Robert E. K. Rourke
James F. Sandford
L. Manlius Sargent, Jr.
Albert P. Shulte
Gerald Tebrow
Paul Tukey
Nancy C. Whitman
Martha Zelinka

Introduction

This book reflects the idea that some knowledge of statistics is important for everybody and that some knowledge of statistics is easily understood by everybody. But, of course, statistics sometimes deals with numbers and averages, with charts, diagrams, and tables. But statistics also asks about events, whether they are unusual, what relationships different events may have, whether past events may help us to predict the course of future ones, and with what confidence such predictions may be made.

The examples in this book introduce and use some of the elementary ideas of probability to help study these and similar problems. We hope these examples will encourage experimenting with such concepts in classrooms. Each example illustrates some statistical idea and will, we hope, serve as an inspiration to both teachers and students for the development of similar but perhaps better examples. After all, no one knows quite as well as the reader just what examples will prove illuminating and exciting. And if the statistical ideas are newer to you than some other mathematical ideas more commonly taught, perhaps these examples will help you to learn in a refreshing and appealing manner.

Contents

Plagiarism and Probability

WILLIAM H. KRUSKAL

INTRODUCTION

"A little knowledge is a dangerous thing." One of the most important things to learn in statistics is how not to misuse it. Here is an example of such misuse. The chances of an admittedly unusual occurrence are made to seem so tiny that a jury might be convinced that the event could not have happened accidentally. We discuss here some of the misconceptions of probability that can lead to such faulty reasoning, and show how to do better.

In The Paper Dragon, a novel by Evan Hunter, the plot turns on a charge of literary plagiarism, and a resulting trial. One of the arguments for the plaintiff in the plagiarism trial was that both literary works were about soldiers in the 105th Division of the U.S. Army; the amazing coincidence that the two authors should both hit on 105 as the number of the Division is given as evidence that the second author (Driscoll) plagiarized the first (Constantine). The following quotation gives a central argument of the plaintiff's lawyer.

"...there were only sixty-seven actual infantry divisions in existence during the time of the Eniwetok campaign, and...when we come to the divisions beyond the designation '100' we

The author is at the University of Chicago, Chicago, Illinois. This example was brought to the author's attention by Mrs. Carolyn H. Bingham.

have the 101st, 102nd, 103rd, 104th, and 106th.
There is no 105th division. Nor was there a
105th division in 1950. There were only seven
actual infantry divisions at that time, the
1st, 2nd, 7th, 24th, 25th, and the 1st Cavalry.
Today there are twelve infantry divisions and,
needless to say, none of them is the 105th,
either.

"Perhaps Mr. Driscoll can adequately explain to this court how he happened to hit
upon those three digits in sequence... . Out
of all the possible numbers Mr. Driscoll could
have used to label his infantry division, he
chose the identical number that appears in Mr.
Constantine's work. This is an amazing coincidence, your Honor, it is almost an impossible
coincidence.

"Now, your Honor, in much the same way
that there are laws governing our society,
there are also laws governing chance, and these
are called the laws of probability, and it is
against these that we must examine this use of
an identical division number. If we were to
take all the digits from zero to nine and try
to figure out all the possible different combinations for any three of those digits, we
would have to raise ten to the third power,
which means we would have to multiply ten times
ten times ten, and that would give us an answer of one thousand possible combinations.
In other words, the odds would be a thousand
to one that any man would choose a specific
combination over any other possible combination. A thousand to one, your Honor. And
those odds, as impressive as they may sound,
are only the odds for a single event. When
we come to two mutually independent events,
the odds are overwhelming.

"What exactly is the probability that
both these men, given the same ten digits,
would then arrange three of them in identical
order? I will tell you, your Honor. The laws
of probability state that in the case of two
mutually independent events, we must multiply
the odds against Event One happening by the
odds against Event Two happening. In other
words, we must multiply a thousand-to-one by
a thousand-to-one, and we then discover that
the odds against Driscoll hitting on this same
combination were a million to one. He had
one chance in a million, your Honor, a de-
plorable cliché to use in a case dealing with
literary matters, but those are the true odds

nonetheless, a <u>million</u> to one, the figures do
not lie.   And even if we wish to give both men
the benefit of the doubt, and say that neither
of them would have designated an Army division
with the number zero-zero-zero--although
stranger things <u>have</u> happened in fiction, as
we well know--even if we were to exclude this
possibility, the odds for both men would be
<u>999</u> to one, and when we multiply that by it-
<u>self</u>, the odds against Driscoll hitting on the
same combination would be 998,001 to one.   A
million-to-one is a neater figure, your Honor,
and will serve our purposes here, I believe.

   "And I believe, too, that with odds such
as these, we are justified in demanding an
explanation, beyond the labeling of such similarities as flimsy and absurd.   Thank you,
your Honor."*

Discuss this argument with special attention to its
assumptions and relevance.   What other probability calculations might alternatively have been made?

<u>Discussion</u>.   There are so many difficulties with the
above over-heated rhetoric that it is hard to know
where to begin a critical discussion.   A good place
might be with the question of ...

   <u>Same number versus specific same number</u>.   The lawyer's argument is that, if each author had independently
and randomly chosen a three digit number (from 000
through 999), then the probability that the two numbers
would both have been 105 is one in a million, 10^{-6}.
   That is correct, but almost surely irrelevant.
One reason for irrelevance is that the computation
should have been of the probability that the authors
chose the <u>same</u> number, whether 105 or some other.
After all, the essence of the plagiarism argument turns
on the identity of the two numbers and not on the particular number 105.
   The probability that both authors chose the same
number--keeping the same assumptions as before--is one
in a thousand, 10^{-3}, a small probability, but much
larger than 10^{-6}.   Exercise 1 asks you to verify this.
   The facts of life here are something like those of
the birthday problem: Ask yourself about the probability that in a group of 25 randomly chosen people at
least two will have their birthdays on the same unspe-
cified day of the year (without reference to year of
birth).   Most people think that the probability is

---

*Evan Hunter, <u>The Paper Dragon</u>, Delacorte Press, New
York, 1966, pp. 202-204.   Reprinted by permission.

small, because 25 is small compared to 365. In fact, the probability is over 0.5, primarily because the particular birthday is not specified--any common date will do--much as allowing any Division number tie above increases the probability 1000-fold.

What about zero? As in the lawyer's speech, 000 should probably be excluded as unrealistic. That changes slightly the probability that the two numbers come out the same under our assumptions. Instead of 1/1000, it is 1/999.

Odds and probabilities. Another minor, but confusing, point is that the lawyer uses the term "odds against" as synonymous with "reciprocal of probability for". For small probabilities that usage is approximately correct, but nonetheless it should be corrected.

To say that the odds are even means that the odds against an event are the same as those for it, and that in turn means that the event has probability 1/2. In general the odds for an event are related to the event's probability. If we denote that probability by p, the odds for the event are $p/(1-p)$. The odds against the event are $(1-p)/p$.

If p is small, $(1-p)/p$ is nearly the same as $1/p$.

In all fairness, we should add that the odds terminology is used loosely and sometimes differently in different circumstances.

What is a reasonable range for Division numbers? Return now to the assumption that the two authors picked numbers from 000 to 999, or maybe from 001 to 999. Keep the assumptions of independence and equal probabilities, but ask yourself whether it makes sense to spread out over so wide a range.

Presumably any author would want to use a Division number roughly in the range of actual numbers, and he would perhaps also wish to avoid a number of a real Division. It is hard to make these notions precise, but if we take the numbering at the time of the Eniwetok campaign, we see that the highest Division number was 106, and that 67 Divisions existed. This leaves 39 unused Division numbers, so that the relevant probability of some number by chance might be as large as 1/39. It might even be much larger, if it were regarded as natural to use a number near the end of the range. For example, if both authors decided to use a number in the 100's, but less than 106, the probability would become 1/2, since only 100 and 105 are unused numbers!

On the other hand, it might be argued that any number up to, say, 150 would be reasonable. Subtracting 67 leaves 83, so from this point of view, the relevant probability would be 1/83. That is small, but much larger than 1/1000.

This discussion shows how easy it is to make misleading computations, and how hard it often is to set up reasonable assumptions as a basis for correct computations. Calculations much like those above have figured in actual court cases, not just fictitious ones.

The assumption of equal probabilities, and a famous mathematical relationship. The above calculations all assume that the numbers in some range are equally likely to be chosen, but that is a very strong assumption. This would be a good point to work on Exercise 2.

On the other hand, the equally likely case gives lower limits for the calculated probabilities. More precisely, suppose that the two authors each guess one Division number out of n possible ones independently and with the same set of probabilities for the various numbers; then the probability that they both guess the _same_ number is greater than 1/n (or equal to 1/n in the case of equal probabilities for the Division numbers).

Let us explore this in the simplest case of two Division numbers. Suppose the guess is of either 101 or 102 with probabilities p_1 and p_2, respectively ($p_1 + p_2 = 1$). The probability that both independent guessers get 101 is p_1^2; that both get 102 is p_2^2. Hence the probability that they both guess the _same_ number is $p_1^2 + p_2^2$. It is convenient to write this as

$$p_1^2 + (1-p_1)^2 = 2p_1^2 - 2p_1 + 1$$

and to regard it as a quadratic in p_1, recalling that p_1 lies between 0 and 1 inclusive. (Note that the choice of 101 and 102 as possible Division numbers was just for illustration; any two would do. We'll discuss the more general case soon.)

You might feel more comfortable with this quadratic function by writing it in conventional symbols as

$$y = 2x^2 - 2x + 1.$$

In Exercise 3 you are asked to prove that the function has its minimum when x (that is p_1) is 1/2, and that the value of the function there is 1/2. The essential point is that the graph of the quadratic function is symmetric about the vertical line x = 1/2.

Further, the value of the function is 1 when x (that is, p_1) is 0 or 1. That makes good sense, for then both guessers are bound to choose the same number.

Thus you have seen that, for $p_1 + p_2 = 1$,

$$p_1^2 + p_2^2 \geq \frac{1}{2} \, ,$$

with equality only when $p_1 = p_2 = 1/2$. A similar result holds in general: for $p_1 + p_2 + \ldots + p_n = 1$,

$$p_1^2 + p_2^2 + \ldots + p_n^2 \geq \frac{1}{n} \, ,$$

with equality only when all the $p_i = 1/n$. (This is a special case of a famous inequality, usually named after one or both of two European mathematicians, Cauchy and Schwarz.)

Hence, when the p_i's are not the same, the calculated probabilities go up, weakening the plaintiff's argument.

In Exercise 4 you are asked to prove, with the help of some hints, the result for $p_1, p_2, \ldots, p_n$.

One might, to be sure, also question the assumption that both authors used the same set of probabilities, but the discussion of that case takes us farther into probability calculations than we want to go here.

Independence. So far we have assumed independence between the two authors in our probability models for choice of Division number. In fact, it is easy to think of sources of dependence, for example, a common friend--or chain of friends--who provided both authors with similar information. We will not, however, speculate further here about models for dependent choice, but it is important to keep in mind that the easy assumption of independence can be seriously in error.

Other sources of similarity. Any two novels about the same military operation are bound to have many similarities just because the backgrounds are similar. It would hardly be surprising if two authors dealing with the same battle area both chose the same city, mountain, or river for part of their respective stories. If numbers come in, Division numbers, serial numbers, unit numbers of enemy groups, telephone numbers, etc., and if both authors guess realistic but not real numbers, the chance that there will be at least one coincidence increases! It is difficult, and perhaps impossible, to make all this precise, but it certainly works in the direction of weakening the plaintiff's argument.

The physicist Luis W. Alvarez, who won a Nobel Prize in 1968, wrote an interesting letter to the journal Science (148, 18 June 1965: 1541) in which he does a probability calculation about the sort of strange coincidence that happens in real life to many of us.

Mr. Alvarez was startled when a newspaper article reminded him of a popular dance band leader he had known about in his student days, 30 years before, and when the same issue of the newspaper carried the obituary notice for the same almost forgotten musician.

So he made a probability calculation to get a rough idea of how often a known person might be recalled in a five-minute period just before learning of his death. Mr. Alvarez assumed that most people know about 3000 other people who might die over a 30-year period. Making some other rough, but serviceable, assumptions, he concluded that the sad coincidence has a probability of about 1/1000 for any one of us over a 30-year period. For 100,000,000 people exposed to the possibility of such coincidences, that means that about 10 coincidences a day will happen. And of course there are many other kinds of coincidences, sad and happy.

Exercises

1. Verify that the probability is 10^{-3} that both authors should choose the same three digit number, if they pick at random and independently from 000, ..., 999.

    [Hint: A simpler case where all possibilities may be written out is that of two well-made dice, independently tossed. For convenience, think of one painted white and the other blue. There are 36 possible ways that numerical values then may appear on the top sides: for example, 1 on both dice, 2 on the white die and 1 on the blue, etc. The probability that both dice show 5 is 1/36, since it corresponds to exactly one of the 36 possibilities, but the probability that both dice show the same numerical value is 6 × (1/36), or 1/6, since there are 6 ways of being the same.]

2. Ask 30 or more people each to name a number between 1 and 10 inclusive. For example, the people might be members of a class--but not the class for which you are studying this set unless it is your individual project only.

    Try to arrange things so that the people choose their numbers without knowledge of other peoples' choices; for example, you might ask everyone to write the number down quickly and quietly on a slip of paper. It is also a good idea to avoid discussion of the purpose of the project, since that might affect choices.

    After you have the numbers, look at the 10 frequencies of the integers to see how much they depart from 1/10 of the number of respondents, that is, from the theoretical expected frequencies if each number were chosen with equal probability. Here is an example from a classroom project of this kind with 32 people.

Integer	1	2	3	4	5	6	7	8	9	10
Number of people choosing it	0	0	5	3	4	3	8	2	5	2

Notice the clustering on 7 and the apparent reluctance to choose the two lowest possible integers.

The usual experience is that integers chosen in this way appear to have different probabilities. People have likes and dislikes for numbers. One way of examining the apparent differences more closely is called the chi-square test, and it is discussed in reference [1].

3. Show that the quadratic function

$$y = 2x^2 - 2x + 1,$$

which may be written

$$y = x^2 + (1-x)^2,$$

has its minimum at $x = 1/2$, and that the value of the function at that minimum is 1/2. Sketch the graph of the function.

[Hint: This can be done by general reference to formulas for the vertex of a parabola, but it is easier and more instructive to notice in this case that the function is symmetric about $x = 1/2$. That is, the value of the function is the same for $x = 1/2 + d$ as for $x = 1/2 - d$, whatever number d might be. The easiest way to see this is by substitution in the $x^2 + (1-x)^2$ form.]

That symmetry suggests that the minimum must be at $x = 1/2$; we know, after all, that there must be a unique minimum since the graph of the function is of a parabola opening upwards. To nail this down, substitute

$$u = x - \frac{1}{2},$$

where u is interpreted as the deviation of x from 1/2. The result is

$$y = (\frac{1}{2} + u)^2 + (\frac{1}{2} - u)^2 = \frac{1}{2} + 2u^2,$$

from which we see immediately both that the function is a minimum when $u = 0$ (i.e., when $x = 1/2$) and that the function's value is $1/2$ at that minimum.

4.   Show that the function of $p_1, p_2, \ldots, p_n$,

$$p_1^2 + p_2^2 + \ldots + p_n^2,$$

where the sum of the p_i's is one, must always be greater than or equal to $1/n$. Show further that the function is equal to $1/n$ only when all the p_i's are $1/n$.

[Hints: First check that you proved a special case of this in Exercise 3, with $n = 2$. That suggests the substitutions

$$u_i = p_i - (1/n),$$

and the exercise then deals with the function of $u_1, \ldots, u_n$, namely,

$$u_1^2 + \ldots + u_n^2 + 2(1/n)(u_1 + \ldots + u_n) + (1/n)$$

under the restriction $u_1 + \ldots + u_n = 0$. The rest of the proof is direct.]

Reference

[1] SBE, Weighing Chances, Set 6.

References labeled SBE are to one of the four parts of Statistics by Example.

Random Digits and Some of Their Uses

ROGER CARLSON

INTRODUCTION

In some sets in this book you are going to be asked to use "random numbers" to help study the problems posed. Since you may not be familiar with this idea, we have included this section to tell what random numbers are and where they come from.

After you have read this explanation, we hope you will understand a bit about what a statistician means when he talks about a random number. You may be interested in how lists or tables of random numbers are developed, so we have given four different methods for making them. Finally, we have given a table of some 2,500 random numbers along with a brief explanation of how to use the table.

Before we explain what random digits are, let's look at an example which uses them. This example involves the <u>simulation</u> of a complex probability problem, and we will need to say something about the technique of simulation.

EXAMPLE. THE MARKSMAN WITH A MEMORY

A certain marksman shoots at a target with a rather small bull's-eye. At first the probability that any one of his shots hits the bull's-eye is 1/2. In other words, his hit-miss sequence is just like the heads-tails sequence of a fair coin. But, unlike a coin, our marksman has a <u>memory</u> and is trying to make a good

The author is at the University of Missouri at Kansas City, Kansas City, Missouri.

score.   Therefore, if he misses the bull's-eye <u>three
times in a row</u>, he aims more carefully, and then his
probability of hitting the bull's-eye is 3/4.   He continues to aim carefully until he finally hits the bull's-
eye.   Then he relaxes and goes back to the old pattern.
In 50 shots, how many bull's-eyes will he hit on the
average?

If the marksman paid no attention to his past
record, we could easily calculate the average number of
hits.   It would be 25; that is, 50(1/2).   But his
memory has made it a very complicated algebraic problem
that we try to solve by <u>simulating</u> his shooting.   We
would proceed as follows: Toss a fair coin and call it
a "hit" when the coin falls heads and a "miss" when it
falls tails.   [P("hit") = P(heads) = 1/2.]   But whenever we get three tails in a row, then we would toss
two coins and score a "miss" if they both fall tails
[P("miss") = P(two tails) = P(tails) × P(tails) =
(1/2) × (1/2) = 1/4] and a "hit" if at least one of the
coins falls heads.   [P("hit") = P(at least one head) =
1 - P(both tails) = 1 - 1/4 = 3/4.]   When we finally
got a "hit" using the two coins we would go back to
tossing one coin.   For our example we would do this for
50 "shots".   The 50 sets of coin tosses would then be-
have just like the hit-miss sequence of the score-con-
scious marksman.   If we toss coins in this way, getting
several sets of 50 "shots" we can then get some idea of
the average number of hits the marksman would make and
the variability of that number from sequence to
sequence.   This procedure--tossing coins rather than
watching the marksman--is called a <u>simulation</u>.   Simulations are frequently used to study complicated problems
in probability and statistics, because we can often run
a simulation fairly quickly and cheaply.   (The same
idea is used when engineers use scale models to check
on calculations.   For example, an engineer who had de-
signed a railroad trestle might test it by running toy
trains over a model trestle.   If the model trestle
breaks, very little is lost and he knows that he better
check on his design.   In our language we would say that
the model trestle and toy train give a simulated test
of the proposed real trestle.)

Exercise

1.   We begin a simulation of the marksman's perfor-
     mance.   We start with one coin and get: THTTT
     (T for tails and H for heads).   Since we have 3
     T's in a row, we toss two coins and get: TT HT.
     This represents a sequence for the marksman of:

     M H M M M M H   (M for "miss", H for "hit").

We are now ready to toss one coin again. Continue the sequence until you get 50 "shots". Calculate the average number of hits for your sequence. Combine your results with those of a classmate to get an improved estimate of this average.

In studying the marksman with a memory we were able to toss coins. But sometimes it is not a good idea to use coins because

a)   it can be a time-consuming job to toss coins (even if it is quicker than watching the marksman) and, in fact, it is practically impossible if we want to simulate a large number of shots, and

b)   if the marksman's probability of hitting the bull's-eye were .527 rather than 1/2 we couldn't use a coin (that is, an ordinary coin) at all.

Both of these difficulties can be overcome by using tables of random digits or by programming a computer to be a random digit generator. So now let's see what a table of random digits is and talk about how such tables are constructed.

GENERATING RANDOM DIGITS

Random digits are digits generated by a process which has the property that each of the possible digits 0, 1, 2, ..., 9 has an equal chance of being the digit selected. (These are random digits base 10. We could have them to any base we want.) Furthermore if the process is really generating random digits, then each selection of a digit is unrelated to the preceding selections. This means that if a process for generating random digits goes on for a while, say generating 1,000 digits then

a)   the frequencies of 0's, 1's, ..., 9's are approximately equal (in this case we would average 100 of each); and,

b)   no procedure based on these 1000 digits will enable us to predict correctly the value of the next digit that will be produced.

In Table 2 we give a page of random digits. You should look at them now. The digits are grouped in pairs so that they look like numbers between 00 and 99. This grouping is just for convenience in reading the table and it is best to think of these as simply pairs of digits rather than numbers from 00 to 99. The pairs are grouped in "blocks" of 25 pairs and there are 50 blocks on a page so that the page has 2 × 25 × 50 = 2500 digits. We recorded 100 digits from the table and

constructed the tally of Table 1. Our tally shows that
we got seven 0's, nine 1's, eleven 2's, and so on.

Table 1. Tally of digits for 100
random numbers

Digit	Tally	Frequency
0	~~1111~~ 11	7
1	~~1111~~ 1111	9
2	~~1111~~ ~~1111~~ 1	11
3	~~1111~~ ~~1111~~ 1111	14
4	~~1111~~ ~~1111~~ 11	12
5	~~1111~~	5
6	~~1111~~ ~~1111~~ 1	11
7	~~1111~~ ~~1111~~ 11	12
8	~~1111~~ 111	8
9	~~1111~~ ~~1111~~ 1	11
Total		100

Exercises

1. Select 100 digits from Table 2 and tally them.
    Combine your tally with the one above and see if
    there aren't about 20 of each digit. If you can,
    combine this with several other students' tallies.
    You can start anywhere in the table, and can "read"
    it from top to bottom or from bottom to top. You
    can read from right to left or left to right. Or
    you can jump around like a knight in chess. By
    changing directions in this way, many more
    sequences of 2500 digits can be found in the table.

2. Now go back to the table and read it digit by
    digit. Cover the next digit with your finger and
    try to predict what it will be before you uncover
    it. Do this for 50 digits and keep a record of
    the success of your predictions.

    Unless you are a wizard (or very lucky) you cannot
    regularly predict the next digit more often than 1
    time in 10. And this is what we meant in (b)
    above.

    To summarize, random digits are digits generated
by a process that ensures that each digit occurs with
about the same frequency as any other digit, and the
next digit is not predictable from inspection of the
previous ones. What processes have these two proper-
ties? Here we describe four methods for obtaining random digits. There are other methods; but, these are
commonly used by statisticians. If you want to start
using random digits immediately, skip on to "Some uses
of random digits".

Table 2.   2500 random digits

	01 02 03 04 05	06 07 08 09 10	11 12 13 14 15	16 17 18 19 20	21 22 23 24 25
01	61 44 34 03 09	05 64 20 54 24	65 69 66 39 80	13 97 76 73 34	41 17 26 81 06
02	85 19 76 44 59	08 60 20 66 68	42 99 28 71 47	73 73 97 24 18	38 25 89 37 20
03	41 17 95 60 40	12 77 51 80 36	46 07 70 39 27	56 65 18 08 70	10 98 45 88 44
04	05 75 79 35 85	60 89 12 66 30	41 25 62 64 54	32 35 64 42 96	16 66 34 64 80
05	36 28 97 63 85	39 85 19 76 44	90 21 61 91 82	73 24 27 32 37	47 98 96 59 75
06	73 56 63 38 40	16 66 09 76 20	08 26 63 16 38	14 16 20 79 87	74 54 96 70 02
07	94 78 81 49 27	50 50 73 46 27	27 04 75 38 83	97 69 20 30 53	19 54 35 62 28
08	28 15 00 92 41	16 37 97 99 37	90 76 87 87 44	34 76 45 94 83	96 07 46 21 97
09	38 37 11 05 75	43 72 07 36 66	43 43 91 05 71	31 59 50 95 94	37 64 64 61 12
10	49 83 04 05 41	49 73 07 47 93	81 28 16 10 73	87 08 07 10 25	01 38 98 14 19
11	85 40 88 30 95	50 02 66 99 44	23 38 99 04 73	69 67 52 23 61	84 89 18 92 79
12	07 55 13 99 88	48 59 19 75 51	96 24 56 41 26	43 14 85 11 47	07 46 03 44 18
13	82 67 78 13 04	83 85 53 97 36	11 63 50 67 81	97 08 14 96 86	82 89 66 78 07
14	83 92 56 64 46	18 94 30 33 99	77 71 38 98 19	31 52 23 35 38	38 55 63 29 61
15	19 60 98 77 85	31 71 71 73 23	08 39 54 26 54	28 61 22 55 59	49 08 28 22 12
16	90 48 51 41 65	32 99 72 09 94	28 43 36 12 21	34 20 08 96 60	96 22 72 84 34
17	90 92 84 99 45	45 79 20 51 76	67 05 60 68 84	10 08 27 22 49	55 93 13 41 34
18	61 28 77 99 36	11 00 00 83 58	83 67 15 23 69	59 60 02 12 94	56 25 84 14 93
19	41 25 77 73 71	53 30 05 60 05	69 30 20 59 30	54 79 23 59 75	78 30 07 29 36
20	58 03 97 70 59	90 92 27 57 38	39 87 40 42 64	44 30 04 33 52	61 53 27 17 60
21	99 60 50 50 60	12 48 08 01 88	60 51 73 38 54	47 41 67 43 84	21 35 12 90 43
22	88 61 29 18 05	31 29 56 94 33	58 88 68 42 10	08 11 15 96 06	15 51 48 39 96
23	19 28 81 63 24	30 96 40 11 59	36 16 01 02 60	36 16 26 83 53	41 09 50 85 01
24	53 61 62 34 47	04 37 74 97 09	77 36 92 80 45	99 26 28 24 24	54 63 43 63 54
25	40 03 44 30 11	42 25 70 19 79	90 12 36 16 80	28 70 24 86 07	76 17 01 50 80
26	65 15 18 13 54	05 13 69 91 51	84 57 52 89 88	12 52 03 39 71	19 48 20 94 16
27	95 79 58 84 86	00 04 73 69 94	89 12 93 84 29	72 62 79 66 98	65 17 54 69 56
28	75 26 86 16 42	65 03 22 43 68	87 68 70 09 18	92 85 94 60 32	97 44 95 82 72
29	92 45 48 29 84	56 60 50 64 07	71 46 35 31 52	21 80 61 25 30	31 99 58 07 04
30	43 00 97 26 90	99 85 55 75 16	09 55 34 16 16	94 32 12 12 07	32 90 97 62 47
31	14 99 59 97 84	18 40 71 98 04	89 24 19 23 56	06 01 68 65 28	23 90 28 10 90
32	79 93 51 89 07	25 25 29 18 02	50 48 21 47 74	61 37 03 51 60	87 97 63 86 43
33	61 37 36 14 84	94 14 96 55 57	05 34 47 88 62	57 73 75 02 34	49 18 83 92 43
34	71 00 51 72 62	59 18 87 82 84	74 04 46 24 66	39 82 30 87 75	41 10 53 02 29
35	76 21 40 24 19	66 19 89 13 48	27 53 41 07 14	28 62 58 67 84	53 16 26 16 97
36	26 97 03 03 30	88 39 46 67 21	17 83 46 74 11	35 54 29 36 86	30 32 06 47 37
37	58 54 81 74 22	32 45 26 40 88	30 91 66 86 52	71 42 99 54 75	12 94 11 09 83
38	45 33 94 97 70	96 27 03 89 63	37 57 46 16 18	78 55 78 07 98	03 46 57 47 39
39	27 26 48 62 10	83 63 45 30 92	48 32 96 67 26	95 90 65 50 46	09 95 58 67 29
40	57 74 80 98 61	50 30 38 41 58	86 28 79 50 71	48 30 58 93 23	70 76 72 42 06
41	68 78 34 95 35	91 63 55 60 22	19 10 77 88 59	11 36 40 56 55	56 29 76 58 93
42	97 63 57 69 62	65 00 51 67 52	21 53 52 16 86	73 67 24 16 68	09 05 74 93 63
43	20 32 35 52 41	47 17 53 83 72	45 20 28 25 04	21 94 00 18 55	26 80 19 80 20
44	53 90 99 23 17	76 44 15 99 65	91 04 22 64 00	39 80 65 21 47	68 75 28 48 16
45	36 42 17 95 78	02 29 66 50 33	65 61 43 77 29	93 34 62 39 42	36 07 61 92 07
46	40 09 18 94 06	62 89 97 10 02	58 63 02 91 44	79 03 55 47 69	14 11 42 33 99
47	33 19 98 40 42	13 73 63 72 59	26 06 08 92 65	63 08 82 45 85	14 45 81 65 21
48	69 49 02 58 44	45 45 19 69 33	51 68 97 99 05	77 54 22 70 97	59 06 64 21 68
49	17 49 43 65 45	04 95 82 76 31	85 53 15 21 70	59 17 27 54 67	07 76 13 95 00
50	43 13 78 80 55	90 80 88 19 13	13 89 11 00 60	41 86 23 07 60	22 77 93 30 83

FOUR METHODS FOR OBTAINING RANDOM DIGITS

Location

One way to obtain random digits is to locate some published table of numbers in the hope that the digits of
these numbers will satisfy the conditions for random
digits. Some places to look would be tables of logarithms, square roots, or other mathematical functions,
telephone directories, almanacs, stock market reports,
tables of physical constants (for example, melting
points of chemical compounds), and so on. But we must
be careful. All such tables are published because
their entries are interesting (to someone), and so it
is likely that the numbers will follow some pattern.
Square roots are a good example of this. But even if
the numbers form a pattern, their digits may not. If
for example we have a 15-place logarithm table, it may
be that the 9th and 13th digits do not form a pattern.
Similarly a telephone directory has to be edited. At
least we may need to remove the telephone numbers of
some businesses since these telephone numbers have many
0's. Even with these precautions phone books can be
biased sources of digits. Table 3 shows the frequency
of 10,000 digits from a phone book. It may not be
obvious but there are too few 9's here for a random
sequence.

Table 3. Frequency of digits from a telephone directory

Digit	0	1	2	3	4	5	6	7	8	9	Total
Frequency	1026	1107	997	966	1075	933	1107	972	964	853	10,000

Several published tables--called "tables of random
digits"--were in fact constructed in this way.
     Finally, almanacs and similar data books record
numbers which tend to be periodic, such as records of
rainfall, and these would not give a very long sequence
of random digits.

Games of chance

Probability theory began as the study of games of
chance; the devices used in these games, such as cards,
dice, coins, and roulette wheels are, therefore, very
naturally still used by statisticians as sources of
random digits. There are three disadvantages to using
games of chance to generate random digits. First of
all it is a slow method for it takes a long time to
shuffle a deck of cards well or to wait for a roulette
wheel to stop. Second, such devices usually don't give
the digits 0, 1, ..., 9 directly. Ordinary dice for
example give the digits 1, 2, ..., 6. (It is possible

to have 20-sided dice, each with 2 sets of the digits
0, 1, ..., 9. To get the digits 0, 1, ..., 9 from such
devices usually involves some waste; for example, ignoring 6 of the numbers on the roulette wheel (and the
0 or 00)--the random digits are just the units of the
numbers which come up. Third, it is difficult to construct (or find) dice or roulette wheels which are
practically unbiased. Casino roulette wheels appear to
develop slight biases after some use and for this reason one statistician (M. G. Kendall) constructed his
own "roulette wheel" in order to prepare a table of
digits which he thought were random.

Random physical processes

Some physical processes seem to be random and one can
generate random digits by monitoring these processes.
Radioactive decay, electrical "noise", and high-fre-
quency oscillators have random aspects to them and have
been used at various times to give random digits. For
example, ERNIE (Electronic Random Number Generator
Indicator Equipment) uses electrical noise in neon
tubes as a primary source of randomness and its output
passes all the tests for random numbers. In principle,
a Geiger counter--if its "clicks" are put in digital
form--is a random digit generator.

The method of pseudo-random digits

With the development of high-speed computers, statisticians and other users of random digits needed large
quantities of random numbers. But such large tables
take up a great deal of computer storage space, leaving
little room for computing. To solve this problem they
developed methods of computing random digits by a formula. At first this appears to be impossible because
if you compute numbers by a formula then you know what
the next one will be and it cannot be a random number.
But to make this prediction requires the formula; you
cannot predict the next digit just by looking at the
last ones. Such formulas are often called pseudo-random digit generators. Pseudo because they are not
strictly random; in particular, after a long series of
digits have been generated they begin to repeat themselves. Most of these formulas are too complicated to
explain here, but if you are a computer user you can
find out, from the manual for your machine, exactly how
these digit generators work. One of these methods can
be described briefly: take a large number like
97021153. Square it, and take the middle eight digits
from the square:

$$(97021153)^2 = 9432518360050419.$$

The middle eight digits are 51836005.

Now square this and take out the middle eight digits of the square and so on. After a large number of such squarings the sequence of middle digits behaves almost like a sequence of random digits.

Throughout our discussion we have mentioned that these processes for generating random digits might not always work. To tell whether a process is really working we need ways of testing for randomness. Some of our later examples deal with such tests. The reader is already wondering why we accepted our first tally (which had only 5 5's when we expected 10) and rejected the table for telephone numbers when there are nearly 1000 of each digit. He'll have to wonder about this for quite a while, these matters are treated in Set 6 of this book and reference [2].

SOME TABLES OF RANDOM DIGITS

There are many published tables of random digits and many computer programs available for generating pseudo-random digits by machine. Here we mention a few of the more readily available tables with a brief description of the process used by their compilers.

1. Tippett, L. H. C., Random Sampling Numbers, Cambridge University Press, 1927. Compiled by editing census figures; contains 41,600 digits.

2. Kendall, M. G. and Smith, B. B., Tables of Random Sampling Numbers, Cambridge University Press, 1939. Gives 100,000 digits obtained from a "roulette" wheel especially constructed for this purpose.

3. Fisher, R. A. and Yates, F., Statistical Tables for Biological Agricultural and Medical Research, Oliver and Boyd, London, England, 1949, Table XXXIII of 6th ed. Compiled from the 15th through 19th digits of a table of logarithms.

4. The RAND Corporation, A Million Random Digits with 100,000 Normal Deviates, The Free Press, Glencoe, Illinois, 1955. "The random digits in this book were produced by rerandomization of a basic table generated by an electronic roulette wheel. Briefly a random frequency pulse source providing on the average about 100,000 pulses per second was gated about once per second by a constant frequency pulse. Pulse standardization circuits passed the pulses through a 5-place binary counter. In principle the machine was a 32-place roulette wheel which made, on the average, about 3000 revolutions per trial and produced one number per second." (p. xi).

SOME USES OF RANDOM DIGITS

Simulating games of chance. By reading a table of random digits properly we can quickly simulate many games
of chance (and often get approximate answers to many
hard problems in probability). Consider first the game
of coin tossing. Since each digit in a random digit
table occurs with frequency 1/10, any 5 digits will
occur with frequency 5/10 = 1/2, and so this should be
like tossing a coin since the chance of heads is 1/2
for a fair coin. All we need to do is read the table
of digits and write down H (for heads) whenever the
digit is even and T (for tails) whenever the digit is
odd. (Zero will be counted as even.) For example, ten
digits from the table are 4 1 1 0 7 6 4 7 9 1 which
gives

     H T T H T H H T T T

for 10 tosses of our simulated coin. It is much quicker
to read the table this way then to actually toss coins
and we don't have to worry about whether the coin is
fair or whether it was tossed fairly.

     Suppose we toss 3 coins; how often do we expect to
get exactly 3 heads? We simulated this problem by
taking 50 triplets of random digits and scoring each
member of the triple H or T as above. Our results are
shown in the table below.

Number of "heads" in 50 tosses of
three "coins" (number of even digits
in 50 triplets from the table)

0	6
1	14
2	20
3	10

     Since 3 heads and 3 tails have the same chance, we
averaged the 6 and 10 to get 8/50 = .16 for the approximate frequency of getting three heads. (The correct
value is 1/8 = .125 but we can't expect to be very
close with only 50 trials.)

     How can we simulate tossing a die? A fair die is
supposed to give each of the numbers 1, 2, ..., 6 with
equal frequency. To simulate die tossing, take random
digits and simply ignore the digits 0, 7, 8, and 9,
using the others for the "die". Thus we took out 10
random digits:

     5 0 1 4 4 9 8 1 0 6

which gave six die tosses with the results:

     5 1 4 4 1 6.

Exercises

1.   Use the table to do the following:

   a)   "Toss" two "dice" until a sum of 7 comes up.

   b)   "Toss" 3 "coins" 100 times, adding your results
        to the result above and again approximate the
        frequency of 3 heads.

   c)   "Draw" cards from a deck without replacement
        until you draw an ace. [Take the two-digit
        numbers 01-52 for the "cards", ignoring the
        others, and let 01, 02, 03, 04 be "aces". But
        also, since you're drawing without replacement
        you must ignore any numbers which have
        appeared previously.]

2.   Drawing a random sample. Suppose we want to
     choose a committee of 3 from a class of 30 stu-
     dents. We want all groups of 3 students to have
     an equal chance of being on the committee--not the
     3 most popular. So we will select them at random
     rather than by vote. (Statisticians would say we
     have a sample of 3 from 30.) This can be done
     quickly with a random number table if we first
     number each student (in any convenient way, say by
     numbering their desks). Then we take 3 different
     numbers from the table, ignoring 00 and numbers
     larger than 30. For example, if we took these
     numbers from the table to get a sample of 3:

     50 14 49 81 06 14 01

     our sample would be the 3 students with numbers
     14, 6, and 1. Note that we ignored the second 14.

     Suppose our 30 students were arranged in 4 rows of
     desks with 8, 8, 7, and 7 students in rows 1, 2,
     3, and 4. Suppose we pick 3 rows at random and
     then pick a student at random from each of these
     rows. Does this give the same results in the long
     run? If you can't reason this problem out, you
     might simulate it.

3.   Using 3-digit random numbers simulate 20 shots for
     a marksman who had no memory and hit the bull's-
     eye with probability .527. Could you do this with
     coins?

4.   Test the table for "memory". Pick any number and
     record 100 digits immediately following 100 occur-
     ences of this number. Is their frequency differ-
     ent from the digits as a whole? [For example, we
     picked 4 and began reading: 93 43 69 64 07 34 18

04 52 44 ... .   The digits immediately following
4's are 3 0 1 5 4 ... .   Continue in this way
until you get 100 digits to see if the 4 affects
their frequencies.]

5.   Table 4 shows the frequency of 1000 digits ob-
      tained from 4 different sources. Do any of these
      show extreme deviation from the expected frequency
      of 100? Why?

Table 4. Frequency distributions of digits from four
sources

Digits	Telephone book	Pseudo-random digits (from a computer)	Table of logarithms	First 1000 digits of the number π
0	110	106	97	90
1	97	99	101	120
2	93	107	102	103
3	92	105	103	98
4	110	100	103	90
5	107	96	101	93
6	102	91	104	95
7	99	103	97	102
8	102	104	97	102
9	88	89	95	107
	1000	1000	1000	1000

6.   Some discussion questions on the idea of simula-
      tion follow. For the discussion use the following
      definition: A is a simulation of B if the out-
      comes of the process A are the same as the out-
      comes of the process B in all essential respects.

      a) Suppose we use a coin to simulate a marksman
          without a memory whose probability of hitting
          a bull's-eye is 1/2. We sit next to him and
          toss our coin while he shoots the gun. A
          strong wind comes up. Is the coin still a
          simulation of the marksman?

      b) To find the length of the face of the Cheops
          pyramid we can do two things: (1) crawl up
          the face with a long tape measure, or (2) mea-
          sure its base and height and calculate the
          length of the face by the Pythagorean theorem.
          Would the results be exactly the same? Why
          (or why not)? Thus, is (2) a simulation of
          (1)? Which method would the Egyptians who
          built it have used? (Or did they use some
          third method?)

c)   Shakespeare said, "All the world's a stage."
Did he mean that drama simulates life?

d)   If you want to know whether you would like to
read a certain book or see a certain movie, you
often ask someone who has read the book or
seen the movie if he liked it. Is his seeing
the movie a simulation of your seeing the
movie? Some people you would not ask; does
this mean that they differ from you in essential respects?

References

[1]   SBE, Weighing Chances, Set 6.

[2]   SBE, Detecting Patterns, Set 5.

How Many Games to Complete a World Series?

MARTHA ZELINKA

INTRODUCTION

Baseball 1971 is now history; the results are recorded.
Long before the playoffs began many people tried to
figure out and predict why their favorite team would
win the 1971 World Series in 4, or maybe 5, or 6, or 7
games. How many were correct in their prediction?
What facts did they consider in making their prediction?

We are going to "play" a lot of games. We will
assume in order to make our playing simple that the
chance of winning a game stays the same for the whole
series. For example, if a team has probability p = .6
of winning, it will have this same probability of
winning for each game of the series. This tells us
also that the other team has probability .4 of winning
each game in the series. Ordinarily the chances of
winning change from game to game.

We know that there are many other factors that
influence the outcome of a game. For example: a team
prefers to play in its own ball park; the conditions
of the field, the weather conditions, injuries, pit-
chers, all influence the outcome of a game. How about
a very close play and the umpire's decision?

We are going to play many series quickly and
peacefully, by simulating each game in the following
way.

The author teaches at the Weston High School, Weston,
Massachusetts.

In Set 2 of this book, you find a table of Random Numbers. A similar table was used in the example that follows; the random numbers in the table came in blocks of ten digits. To study a situation where the teams are unevenly matched, we assign the digits 1, 2, 3 to the American League team, the other digits, 4, 5, 6, 7, 8, 9, 0 to the National League team.

Thus, if the entry used is

   5 2 4 3 8 2 5 8 4 0,

the series would look like this:

   5 2 4 3 8 2 5 8 4 0
   ↓ ↓ ↓ ↓ ↓ ↓ ↓
   N A N A N A N

and the National League would have won in seven games. If the entry is

   8 7 0 0 1 8 4 2 7 3

the series would look like this:

   8 7 0 0 1 8 4 2 7 3
   ↓ ↓ ↓ ↓
   N N N N

and the National League won in four games.

Exercises (Discussion)

1.   In the series that we are going to simulate we will assign 3 of 10 digits to the American League and 7 of 10 digits to the National League as above. What is the probability, in each game, that the National League will win? That the American League will win?

2.   How many of the digits in each block of 10 do we need to simulate a series?

3.   To win the World Series a team must win 4 out of 7 games. Who wins the last game of the series?

4.   If a team is to win in 6 games how many must the team win of the first 5 games?

5.   The American League won the World Series in 1949 in 5 games, the score was 4 - 1. The series could have gone AAAN A, or ANAA A. Find the other ways in which the series could have been won in 5 games.

EXAMPLE 1

Now to our simplified World Series, played with random digits as described earlier. We assume that the National League team is the stronger team and has probability p = .7 of winning each game. We want to know: What is the mean number (average) of games required to complete the series? To approximate the answer to this

question we simulate a hundred series and record the
results, assigning digits 1, 2, 3 to the American
League; 4, 5, 6, 7, 8, 9, 0 to the National League.
The results are shown in Table 1.

Table 1. Simulation of 100 series: p = .7. The series
is won in 4, 5, 6, or 7 games.

```
       ↓ ↓ ↓ ↓         ↓ ↓ ↓ ↓         ↓ ↓ ↓ ↓         ↓ ↓ ↓ ↓
1 2 3 4 5 6 7   1 2 3 4 5 6 7   1 2 3 4 5 6 7   1 2 3 4 5 6 7

N N N N          A N N N N        A N N N N        N A A N N
A N N N N        A N N N N        N N A A N N      N N A N N
A N A N A N N    N N N N          A A N N A A      N N A A N N
N A N A A N A    A A N N N A N    N N N N          N N N N
N N N A N        A N N A N N      N N N A N        N A A A N N N

N N A N N        N N N N          N A N N A N      A N N N N
N A N N A A A    A A N A N N A    N A N N N        N N A N N
N N N N          N N N N          A N A N N N      A N A N N A A
N N N A A N      N N N N          A A A N N A      N N N N
A N N N          N N N N          A A N A N N N    N N N A A N

N A N A A N A    A N N N A N      N N N N          N N A N A N
N A N N N        N N N N          A A A N N A      A A N N N
N A N N          A A N N A N N    N N N N          N N N A N
N A A N A A      N N N N          A N N N A N      N A A N A N A
N N A A N N      N N N N          N A N A A A      N A A N A A

N N A A N N      N N N N          N N N N          A N N N N
N N N N          N A N N A N      N N N N          N N A A N A N
N N A N N        N N A N N        N N N N          A N A N A N A
N N N N          N N A N A A N    N A N N N        N A A A N N A
N N N A N        A N A N N N      N A N N A N      N N A N N

A N A A N N N    N A N N N        N N A N A N      N N A N N
N N N N          N N N N          N N N A N        N A A N N N
N N A N N        N N A N N        A N A N N A A    N A N N N
N N N N          A N N N          N N A N N        N A N N N
A N N N          N N A N N        A N N N N        A N N N N
```

Frequency distribution for completing the series in

Number of games	4	5	6	7	Total number of games
Frequency	~~1111~~ ~~1111~~ ~~1111~~ ~~1111~~ 111	~~1111~~ ~~1111~~ ~~1111~~ ~~1111~~ ~~1111~~ ~~1111~~ 1111	~~1111~~ ~~1111~~ ~~1111~~ ~~1111~~ ~~1111~~ 1	~~1111~~ ~~1111~~ ~~1111~~ 11	537
					Total series
	23	34	26	17	100

From the frequency distribution for these 100
series we can now compute M, the mean number of games
it took to complete a series:

$$M = \frac{4(23) + 5(34) + 6(26) + 7(17)}{100}$$

$$= 0.01(92 + 170 + 156 + 119) = 5.37$$

Exercises for Example 1

1.   Check the frequency distribution and the computa-
     tion for M = 5.37, the mean number of games it
     takes to complete the series.

2.   Find the average percentage of games won by the
     winning team of the series; the percentage of
     games won by the losing team.

3.   Let a group of students (all readers are students)
     work together in pairs. One member reads the
     digits of Random Numbers (you could use the table
     of Random Digits in Set 2 of this book), the other
     person translates into A or N and records.

     a) Use p = .7 as in Example 1, and after 100
         simulations compute the mean number M. Com-
         pare the result of the group effort with the
         average for 100 series in Example 1.

     b) Repeat with p = .6; with p = .5; with p = .9.

     c) Use the data to make a graph of the mean num-
         ber of games against p as p goes from .5 to
         1. Sketch a smooth curve through the points.
         Use it to estimate the mean number of games
         when p = 0.75.

     Note: When you draw a graph you must decide which
     variable should be represented on the horizontal
     axis, which on the vertical axis. The mean number
     of games increases or decreases, goes up or down,
     as p changes. So to draw a graph "against p"
     means the values of p are marked off on the hori-
     zontal axis.

4.   Look up the statistics for World Series records
     from 1950 - 1969, compute the average number of
     games. Compare this result with those of the
     simulated series.

5.   In the 1970 Playoffs

         Baltimore won over Minnesota
         3 - 0,
         Cincinnati won over Pittsburgh
         3 - 0.

In the 1971 Playoffs

> Baltimore won over Oakland
> 3 - 0,
> Pittsburgh won over San Francisco
> 3 - 1.

a) Discuss the results of these "3 out of 5" series. These results are unusual.

b) Simulate 100 playoff series; assign probabilities of your choice. Do you want the teams well matched? If so, what value of p should you use? Find a partner, so that you can play series with different values for p.

c) For each different value of p compute the mean number of games required to determine the Pennant Winner in (b).

d) Compare the 1970, 1971 facts with your results and discuss.

On pages 29-31 are Tables 2, 3, and 4, which together with Table 1 are used in Exercise 6.

6. We can make many observations from the summary in Table 5. First, however, check all entries and computation against the recorded simulations, then answer the following questions.

a) What proportion of the series end in 4, 5, 6, or 7 games?

b) Compare the proportion of series ending in 4 games to those ending in 7 games as p changes from .5 to .6, to .7, to .8.

c) Study the following graphs. Explain how each point was plotted. What is the relation between questions (a) and (b) and the graphs?

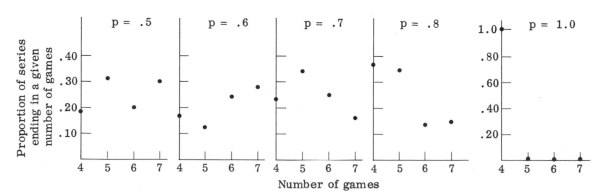

d) Use the data to draw a graph of the mean number of games as p goes from 0 to 1.0. From the graph, estimate the mean number of games for p = .9.

e) From a simulation of 100 series with p = .9 the following frequency distribution was recorded:

Series was completed in

Number of games	4	5	6	7
Frequency		60	29	10	1

Compute the mean number of games needed to complete the series. How close is this mean to your estimate from your graph in (d)? Enter the point in your graph in (d). Compare the results with those of Exercise 3, where you used your own simulations and data.

f) Use the data to draw a graph of the probability that the better team wins, P(better team wins), as p goes from 0.5 to 1.0.

g) Use the data to draw a graph of the probability of winning the series, P(win series), against p, for p from 0 to 1.0.

h) Use the data to draw a graph showing the number of winning games against p, as p goes from .5 to 1.0 (i) for the American League, and (ii) for the National League.

Table 2. Simulation of 100 series: <u>p = .5</u>. Assign
digits 1,2,3,4, and 5 to the American League, 6,7,8,9,
and 0 to the National League. This means that p = .5
for each team, but "<u>your</u>" team is better!

```
N N N A A A A A   A N N A N A A   A N A A N A     A N A N A N N
N N N A N         A N A N A A     A A A N A       N N A N N
N N N N           A A N A A       A A A A         A N A N A A
A N A N N N       N N N A A N     N A N N N       A N A A A
A A A A           A A A A         N A N N A N     N A A N N A N

N N N N           N N N A N       N A A A A       N N N N
N N A A A N A     A N N N A A A   N N A N N       N A N N N
A N A A A         A A N A A       N A A A N N A   A A A A
N A N N A A N     A A A A         A A A A         A N A A N N A
N N A A N A N     A N A A N A     N N A N N       N N N N

N N N N           A A N N A A     A N N N N       A N N A A A
A N A A A         A N A N A N N   N A A N N A A   A N A N N N
N A A A N A       A A N A A       A A N A A       N N A N A A N
A A N N A N N     A A A A         N N N A N       N N A N N
N A A N N A A     N N N A N       A N N N N       N A A A N N N

N A N A A N N     A N N A A N A   N N A A N N     A A A A
A A N N N N       A N A N N N     N N N N         N N N A N
N N A A N A A     A A N N N N     A A A A         A N A A A
N N A N A A A     A N A A A       N A A N N A A   A A N A N N A
N N A A A A       A N N N A N     A N N A A N A   A A A A

N N A A N N       A N A A A       N A N A N A A   A A N A A
A A A A           N N N N         A N A A N N N   A N N A A N A
A A A N A         A A A N A       A A N N N N     N N A A A A
N N N A N         N A A A N N N   A A A N A       N N A N N
A A A A           N A A A N N N   A N N A A A     N A N N A A N
```

Frequency distribution for winning the series

Number of games	4	5	6	7	
Frequency		19	31	20	30	M = 5.61

Table 3.   Simulation of 100 series: <u>p = .6</u>.   Assign digits 1,2,3, and 4 to the American League, 5,6,7,8,9, and 0 to the National League.

```
A N N N N     N N N N       N A N N A A N  A N A N N
N N N A N     N A N N N     N A N A N A N  N A A N A N N
A N N A A N A A A N N N N   N N N A N       A N N A N N
A A A A       N N N N       A A A N A       A N A A N N
A A N N A A   N A N N N     N N A N A N     N A N A N N

A A A N A     N N A A N N   A N A N N       A N N A N A N
N N N N       A N N A N A N A A A A         N A N N A N
N N N N       N A A N A A   N N N N         A A N N N A N
N N A N N     N A N A N N   N N N A N       N N N N
N A A N N N   N N A N N     N A N A N A A   N A A A N N N

A N N A N N   A N A A A     N A A N N N N N N A N N
A N N A N A N N N N N N     N A N N         N N A N A A A
A N N N A A A N A A N N     N A N A N       N N N N
N A A A A     N N N N       A N A A         A N N N
A A N N N N N N A A A N A   N N A A N N     N N A A A A

N N N N       N A N A A N A A A A N N N N   A A A N N A
A N A A A     N A A N A N N N A A N A N N   N A N N N
A N A N A N A A A N A A     A N N N         N A N N
A N A A N N A A A A N A     N A N N N       A N A N N
A A N N N N   A N N N N     N A N N N       N N N A A N

A N N N N     A N N N N     N N N N         A N A N N
A N A A N N A N A N A A N A A N N A N A N  A N A N A N A
A A N A A     N A A A A     N N N A N       N A A A N N A
N N N N       A N N N       N N N N         N A N A N A A
N N N N       N N N N       N N N N         A N N A N A N
```

Frequency distribution for winning the series

Number of games	4	5	6	7	
Frequency	17	31	24	28	M = 5.63

Table 4. Simulation of 100 series: p = .8. Assign digits 1 and 2 to the American League, 3,4,5,6,7,8,9 and 0 to the National League.

```
N N N A A A N | N N N N        | A N N A N A N | A N N N N
N N N N       | A N N N N      | N N N N       | N N A N N
N N N N       | A N N N A N    | A A N N N A N | N N A N N
A N A N N N   | N N N A N      | N A N N       | N N A N A N
A N A A N N A | N A A N N N    | N N N N       | N A A N N A N

N N N N       | N N N N        | N A A A N N N | N N N N
N N A A N N   | N N N N        | N N N N       | N A N N
A N N N N     | N A N N N      | N A A N N N   | A N N A N N
N A N N N     | A N A N N N    | A N N N N     | A N N N N
N N A N N     | N N A N N      | N N N N       | N N N N

N N N N       | N A N N N      | N N N N       | A N N N A A A
N N N N       | A N N N N      | N A N N N     | N N A N N
N A N N N     | N N N A A A N  | A N N A A N N | N N N N
A N N N N     | N N A A N N    | N N N N       | N N N N
N A A N N A N | N N N N        | N N N N       | N N N A N

N A N N N     | A N N N N      | N N A N N     | N N N N
N N N N       | N N N N        | N N N N       | N N N A N
N N A N N     | A N N N N      | N N N N       | N N A A N N
N N N N       | A N N A A N A N| N A A N N N   | N A N N N
N N N N       | N N N N        | A N N A A N N | N N N N

N N N N       | N N N N        | N N N N       | A A N A N N N
N N N N       | N N N N        | N N N N       | N N N A N
N A N N N     | N N A N N      | N A N N N     | N N A A A N N
N N N N       | N A N N N      | N N N N       | N N A N N
N A N A N N   | N A N A N N    | N N N A A A N | N A N N N
```

Frequency distribution for winning the series

Number of games	4	5	6	7	M = 5.06
Frequency	37	35	13	15	

Table 5. A summary of four sets of 100 simulated World
Series, with p = .5, p = .6, p = .7, p = .8, q = 1-p,
where p = P(National League wins game).

p	q	Games to complete series				American League wins		Mean number of games to end series
		4	5	6	7	No. of series	Proportion of games	
.5	.5	19	31	20	30	54	$\frac{292}{561}$ = .53	5.61
.6	.4	17	13	24	28	29	$\frac{222}{563}$ = .39	5.63
.7	.3	23	34	26	17	15	$\frac{158}{537}$ = .29	5.37
.8	.2	37	35	13	15	3	$\frac{109}{506}$ = .22	5.06

(Frequencies)

Classifying Pebbles, a look at the Binomial Distribution

RICHARD F. LINK
AND MICHAEL L. BROWN

INTRODUCTION

Geologists sometimes want to know the composition of
pebbles in a stream. In one method of classification
called binomial, the pebbles are identified as either
being of a given type, such as quartzite, or not being
of that type. For example, a geologist might wish to
find the fraction of pebbles in a stream which are
quartzite. Quartzite is an easily recognized mineral.
If we assign the number 1 to quartzite pebbles and 0 to
nonquartzite pebbles, we can see what totals we get
when we repeatedly pick up samples of 10 pebbles from a
stream and classify them.

AN EXPERIMENT IN PICKING PEBBLES

If the pebbles have been randomly distributed in the
stream and if the geologist is not influenced by unusual characteristics of the pebbles, as would happen if
he picks up only bright, shiny pebbles and ignores all
others, then the process of picking up pebbles and
making a binomial classification should generate data
which follow a binomial distribution. We want to describe this distribution and illustrate it for this problem. If we repeat this experiment 100 times we will
observe a distribution of X, the number of quartzite
pebbles in a sample of 10 pebbles. Such an experiment

Richard Link is at Artronic Information Systems, Inc.,
New York, New York and Princeton University, Princeton,
New Jersey and Michael Brown is at Harvard University,
Cambridge, Massachusetts.

was conducted in 1967 by R. Flemal, in a section of the Gros Ventre River in Wyoming. Table 1 shows the results.

Table 1. Frequency distribution
of number of quartzite pebbles
in samples of 10 pebbles

Number of quartzite pebbles in sample	Frequency f
10	6
9	25
8	31
7	28
6	9
5	0
4	1
3	0
2	0
1	0
0	0

100 = sum of f

If we consider the act of picking up a pebble and classifying it as conducting one binomial trial, the data of Table 1 illustrate the outcome of 100 binomial experiments each having 10 trials. The results of such an experiment could be predicted from statistical theory if the assumption about the randomness and independence of the trials is correct, and if the chance that a pebble is quartzite is known.

CALCULATING THE CHANCE OF GETTING A QUARTZITE PEBBLE

We first use the data in Table 1 to calculate the chance of a pebble being quartzite, and then we develop a formula for the binomial distribution and compare the results predicted by the formula with those actually obtained by Flemal.

We may calculate from Table 1 that, of the 1,000 pebbles examined, 787 were quartzite. Thus, the chance of a pebble being classified as quartzite may be estimated to be 0.787 = 787/1,000. If we had conducted more or fewer trials we would have gotten a slightly different estimate of the chance that a pebble was quartzite, but this estimate should be reasonably close to the "true chance" for this section of river.

A FORMULA FOR THE BINOMIAL DISTRIBUTION

We will now consider the problem of obtaining the formula for the binomial distribution. The derivation will be in two parts. We first calculate the probability of obtaining a particular sequence of 0's and 1's. Then we will find how many different sequences give the same value of X, the number of 1's obtained. By putting these two results together, we get the desired formula.

There are 2^n ways of writing down sequences of n 0's and 1's; that is, there are 2^n distinct sequences of outcomes for n binomial trials. For n = 1 there are two outcomes, 0 or 1; for n = 2 for each outcome for the first item there are two outcomes for the second item, or four different outcomes; see Table 2.

Table 2.   Possible outcomes of 2 binomial trials

Outcome of 1st trial	Outcome of 2nd trial	Combined outcome of both trials		Number of 1's (X)
0	0	0	0	0
0	1	0	1	1
1	0	1	0	1
1	1	1	1	2

Continuing this argument, we see that for 3 items there are 8 outcomes, which are listed in Table 3. Thus, we see the number of outcomes increases as follows: for n = 1 the number of outcomes is $2 = 2^1$, for n = 2 the number of outcomes is $4 = 2^2$, and with n = 3 the number of outcomes is $8 = 2^3$. We can argue inductively that (n-1) trials generate 2^{n-1} distinct sequences of 0's and 1's or 2^{n-1} distinct outcomes, and for each of these the nth item generates 2 more, giving 2^n distinct outcomes in all.

Let us note that, even when the chance of getting a 0 or 1 is equal, this does not mean that each value of X (the sum of the 1's) has the same chance when the sample size is 2 or more. Note that for n = 2 there are 2 ways to get X = 1, and only one way each to get X = 0 or X = 2. For n = 3, we find the results given in Table 4.

Table 3.   Possible outcomes
of 3 binomial trials

Combined outcomes	Number of 1's (X)
0 0 0	0
1 0 0	1
0 1 0	1
0 0 1	1
1 1 0	2
1 0 1	2
0 1 1	2
1 1 1	3

Table 4.   Distribution of number of 1's for
3 binomial trials with p = .5

X	f	Chance (p = q = .5)
0	1	1/8
1	3	3/8
2	3	3/8
3	1	1/8

Total number
of outcomes   =   8   1 = Total probability

More generally, if the chance of getting a 1 is p
at each trial and if the chance of getting a 0 is
q (= 1-p), all outcomes do not have the same chance of
occurring. This case for n = 2 is illustrated in Table 5.

Table 5.   Distribution of number of 1's (X) for 2
binomial trials with arbitrary p

1st trial		2nd trial				
Outcome	Probability	Outcome	Probability	Combined outcome	Combined probability	X
0	q	0	q	0 0	q^2	0
		1	p	0 1	qp	1
1	p	0	q	1 0	pq	1
		1	p	1 1	p^2	2

X	Probability
0	q^2
1	2pq
2	p^2

We assume that the successive trials are independent. If p = 0.6, then 0.6 of "first trials" will give a 1. Now 0.6 of these are followed by 1's and so the proportion of times two trials both give 1 is 0.6 × 0.6 = 0.36. Similarly, the chance of getting two 0's is 0.4 × 0.4 = 0.16.

Mathematically, the assumption of independence between two events A and B means, by definition, that the probability that both events occur is the product of their respective probabilities. That is,

P(A and B occur) = P(A occurs) × P(B occurs).

In our example, define the events A and B by

   A = {1 occurs on first trial},

   B = {1 occurs on second trial}.

Then if A and B are independent, and the probability of a 1 is 0.6, the probability of the sequence 11 is 0.6 × 0.6 = 0.36 as we computed on an intuitive basis above.

Similarly, the probability of the sequence 101 for 3 successive independent trials would be $0.6 \times 0.4 \times 0.6 = 0.144 = (0.6)^2 \times (0.4)$.

We may further illustrate these calculations by again considering the case of n = 3. There are still $8 = 2^3$ possible outcomes, which are listed along with the chance that each occurs in Table 6.

Table 6.   Possible results of 3 binomial trials

Sequence	Chance of occurrence (Chance of 1 equals p, of 0 equals q)	X
0 0 0	$q \cdot q \cdot q = q^3$	0
1 0 0	$p \cdot q \cdot q = pq^2$	1
0 1 0	$q \cdot p \cdot q = pq^2$	1
0 0 1	$q \cdot q \cdot p = pq^2$	1
1 1 0	$p \cdot p \cdot q = p^2 q$	2
1 0 1	$p \cdot q \cdot p = p^2 q$	2
0 1 1	$q \cdot p \cdot p = p^2 q$	2
1 1 1	$p \cdot p \cdot p = p^3$	3

We see from Table 6 that the chance of a particular sequence of outcomes occurring depends only on X and p and that this chance is $p^X q^{3-X}$, where p is the chance that a trial will yield a 1, q the chance it will yield 0, X is the total number of 1's, and 3 is the number of trials. More generally, if we have n trials with X 1's, the probability of a particular sequence is $p^X q^{n-X}$.

We next calculate how many sequences of 0's and 1's have the same value of X.

This can be done by an inductive argument as follows:

	Outcome	Number of outcomes
1st trial	0 1	2
2nd trial	0 1 0 1	4
3rd trial	0 1 0 1 0 1 0 1	8
X	0 1 1 2 1 2 2 3	

We may trace the chain of events which produces a particular X. From the above we see that with 2 trials there is one path to X = 0, two paths to X = 1, and one path to X = 2. Similarly with 3 trials, one path yields X = 0, one yields X = 3, while 3 paths yield X = 1 and 3 others yield X = 2.

We can follow this path approach systematically by starting with n = 1, which can produce an X of 0 or 1. Then for another trial there is only 1 path to X = 0, 2 to X = 1, and 1 to X = 2. Note that, in Table 7, the number on a line is the sum of the two numbers immediately to the right and left of it on the line above.

Table 7.   Number of ways to obtain X

Number of trials	Pascal's Triangle
1	1 1
2	1 2 1
3	1 3 3 1
4	1 4 6 4 1
5	1 5 10 10 5 1
6	1 6 15 20 15 6 1

Thus, to get line 3 from line 2, we have:

— + 1 = 1,   1 + 2 = 3,   2 + 1 = 3,   1 + — = 1,

going from the possible X for 2 trials (note X is excluded from the table for simplicity) to X for 3 trials.

Another way to see how the pattern of Table 7 develops, indeed to show that it works, is to take the pattern 1,3,3,1 which holds for 3 trials. What is the pattern for 4? Each of the counts 0,1,2,3 could either be unchanged with the addition of a fourth trial or be increased by 1. Thus the 16 ways of obtaining X for 4 trials could be presented as follows:

		Total X				
		0	1	2	3	4
If new trial adds	1		1	3	3	1
	0	1	3	3	1	
Number of ways to obtain X		1	4	6	4	1

Similarly, adding a fifth trial would give the following:

		Total X					
		0	1	2	3	4	5
If new trial adds	1		1	4	6	4	1
	0	1	4	6	4	1	
Number of ways to obtain X		1	5	10	10	5	1

The array displayed in Table 7 is known as Pascal's Triangle. These numbers may also be generated by the following formula:

$$\text{(Number of ways } X = c) = \frac{n!}{c!(n-c)!} ,$$

where n is the number of trials examined and

$$n! = n(n-1)(n-2) \cdots 2 \cdot 1.$$

It is conventional to define $0! = 1$. Thus if $n = 5$, and $X = c = 2$ we have

$$\frac{5 \cdot 4 \cdot 3 \cdot 2 \cdot 1}{(2 \cdot 1)(3 \cdot 2 \cdot 1)} = 10.$$

Combining these two results, we have our binomial formula. Given n independent binomial trials, the chance that X = c (where the chance of getting a l is p) is

$$\frac{n!}{c!(n-c)!} \, p^c q^{n-c} \, .$$

COMPARING CALCULATIONS WITH FLEMAL'S DATA

We may use this formula to construct a theoretical model for the results that Flemal might have expected to obtain assuming that the probability of getting a quartzite pebble is .787.

Using our binomial formula, we can calculate the right-hand column of Table 8.

Table 8. Frequency distribution of number of quartzite pebbles in samples of 10 pebbles calculated from binomial formula (p = .787)

X	Observed count	Calculated frequency (%)
10	6	9.1
9	25	24.7
8	31	30.0
7	28	21.7
6	9	10.3
5	0	3.3
4	1	0.8
3	0	0.1
2	0	0.0
1	0	0.0
0	0	0.0
		100.0

From Table 8 we see that we would expect about 4 sets of ten trials to have X = 5 or less out of 100 sets of ten trials.

Table 8 shows that the agreement between the calculation using the formula and Flemal's experimental results is quite close. This means that the assumptions that the distribution of the pebbles in the stream is random, and that their choice by the geologist was impartial, are apparently correct.

Exercises

1. Extend Pascal's Triangle (Table 7) for 7, 8, 9, and 10 trials.

2.   Add the numbers in each row of Pascal's Triangle. What should this sum be? [Hint: The Triangle counts the number of ways each of the different values of X can occur.]

3.   Calculate, using the binomial formula, the probabilities in Table 8 for n = 10, X = 9. (For ease of calculation, use p = .8.)

The following set of data records the results of an experiment comparing the amount of sleep obtained using a new sleeping compound. Assign a 1 to those who gained in sleep using the compound and a 0 to the others.

Subject	Hours of sleep gained using compound
1	1.9
2	0.8
3	1.1
4	0.1
5	-0.1
6	4.4
7	5.5
8	1.6
9	4.6
10	3.4

4.   What is your estimate of p, the probability of gaining sleep?

5.   If p = .5, what is the chance of getting exactly 9 gains and 1 loss, assuming that the binomial formula applies to these data?

6.   Calculate, using the binomial distribution with p = .5, the chance of getting a value of X equal to 3, 4, 5, 6, or 7, in 10 tries.

7.   In the experiment just described, what would we do to find out whether the medication had the same effect on men and women? Describe how you might do the experiment and analyze the results.

8.   These data might not be appropriately analyzed using the binomial distribution. Why?

9.   What two numbers completely determine a binomial distribution?

10.   Use the data of Table 1 to verify that 787 quartzite pebbles were observed.

11.   The reason there are 2^n sequences of n 0's and 1's is that, for each trial in the sequence, we have

exactly 2 choices: 0 or 1. Now suppose we have 3 choices: let's call them A, B, C; it doesn't matter what we label them, since we're just counting the number of patterns they form. How many sequences of length n = 2 can we now make? How many of length n for an arbitrary n? How many sequences of length n can we form if, on trial i, we have c_i possible choices (i = 1,...,n)?

12. Verify the law of construction of Pascal's Triangle, that is, prove

$$\frac{n!}{(X-1)!(n-X+1)!} + \frac{n!}{X!(n-X)!} = \frac{(n+1)!}{X!(n-X+1)!} .$$

13. The average value of X is by definition

$$\mu = 0P(X=0) + 1P(X=1) + ... + nP(X=n).$$

Use the results of Table 5 to show that the average value of X is 2p.

14. One of the features of the binomial distribution which accounts for its popularity is its incredibly wide applicability. Quite literally, looking at occurrences as binomial distributions is as easy as yes and no! Very many events which can be looked at as happening in exactly two ways (yes, no; do, don't do; "black, white") might give rise to a binomial distribution, if only we're a little bit careful. That's what we did in the exercise above about the sleeping compound: "gained, didn't gain". But we were careful to use the same sleeping compound so that the probability of gaining sleep was the same for each trial (person). We also used a different person for each trial, so that trials were independent.

There are undoubtedly many binomial experiments in your life. In this and the next exercise, you're going to construct one. First think of something you do on some days, but not on others. Try to think of something which you'd be as likely to do on any one day as on any other (so its probability remains constant), and the chance of doing it on any particular day doesn't depend on whether you did it or planned to do it on any other day. Activities with a periodic quality won't do; you probably won't shop today if you did it yesterday, so both constancy and independence go out the window. Arguing with your little brother? That's getting warmer, but if doing that yesterday makes you avoid him today (so that Probability (argument one day after a fight) = 0), that's out too. So it is somewhat tricky after all; not all real-life situations are as "neat" as picking pebbles. Very often the mere fact that people are involved makes a randomness assumption very

suspect. Even in the pebbles situation, we were somewhat surprised to find that the geologist actually <u>did</u>
seem to be choosing his rocks impartially.

Let's wait until the next exercise to consider
your own activity. For now, let's try to examine constancy and independence in a true-to-life (albeit
tongue-in-cheek) situation somewhat "messier" than our
geologist's. (Words like "neat" and "messy" refer
loosely to how tricky it is to take account of all factors in a situation which may affect the conclusions of
your model.) Suppose you read the morning newspaper
whenever the dog hasn't chewed it up first; that's a
promising example since it shifts the responsibility
for constancy and independence to the dog, who isn't
nearly as likely to be burdened with nonrandom worries
like periodicity as you are.

Invent three reasons why you can't trust the dog
either. The first should violate both constancy and
independence; the second, constancy but not independence; and the third, independence but not constancy.

15. If worrying about our dog has made you skeptical
     about your initial choice of activity, you might
     try thinking of something you're supposed to do
     every day but sometimes miss--like brushing your
     teeth.

     Let "A" be a symbol for "your chosen activity on
     any particular day".

     a) Estimate p = the probability of doing A.

     b) What is the probability that you won't miss A
         even once in the coming week? that you'll
         miss exactly once? at least once? (How are
         the answers to the first and third questions
         related? Answer this before putting pen to
         paper.)

     c) What is the probability that you won't do A
         even once in the coming week? that you'll do
         it exactly once? at least once?

     d) Answer parts (b) and (c) for p = 0; p = 1.
         Is there a value of p for which you would ex-
         pect the answers to parts (b) and (c) to be
         the same? Describe in words, not symbols, why
         the value you choose makes sense. Although he
         sometimes doesn't bother to tell you, every
         good scientist is in the habit of reasoning
         this way; by examining the way a situation be-
         haves at the extremes, he almost invariably
         gains insight into what "makes it tick" in
         between.

e) The generalization of Exercise 10 to arbitrary
    n is: average value of X = np. One way to
    interpret this average value for the binomial
    distribution is to say: in n trials the <u>ex-
    pected value</u> of X is np. We mean that as we
    take more and more trials (days, in this
    exercise), it becomes more and more unlikely
    that X differs by a large percentage from np.
    Why should the value np come as no surprise?
    How many days would you expect to have to wait
    until you've done A at least 5 times? 10
    times? 100 times? 1,000 times? At what
    value of n do you think constancy or independ-
    ence begins to be doubtful?

Black and White Survival in the United States

BRADLEY EFRON

INTRODUCTION

From the statistical tables of death rates at different ages for black and white men and women, we can extract a great deal of information about the life expectancies of various groups of people. A careful study of these tables, as illustrated in this set, will enable a student to acquire a good knowledge of what simple probabilities mean, both for populations (that is, what proportion will be living a year from now?) and for individuals (what is my chance of dying in the next year?).

The table on page 46 gives the probability that a newborn child in the United States will live to* a given age.

The questions and answers which follow show how elementary probability theory can be used to get information from the table.

EXAMPLE 1

Among 1000 white male babies, how many can we expect to be alive on their 55th birthday?

The author is at Stanford University, Stanford, California.

*"to" means "at least to" in this problem. The data is from the U.S. Census of 1960. "Black" actually means "non-white", and therefore includes groups other than blacks. The main idea is to compare survival in minority and majority groups.

1

Age	White male	White female	Black male	Black female
1	.974	.980	.951	.961
5	.970	.977	.943	.955
10	.967	.975	.939	.952
15	.965	.974	.935	.949
20	.959	.971	.920	.945
25	.951	.968	.914	.938
30	.944	.965	.897	.928
35	.936	.960	.878	.912
40	.924	.953	.849	.892
45	.905	.942	.812	.864
50	.874	.925	.766	.828
55	.824	.900	.701	.783
60	.755	.863	.626	.721
65	.658	.807	.527	.641
70	.538	.725	.418	.539
75	.402	.607	.291	.406
80	.261	.448	.189	.288
85	.131	.262	.096	.174

Solution.   824.   That is what the table means. (Of course 824 is only an average number, and in any particular group of 1000 white male babies the number of survivors will probably differ from the average by a small amount.)   Let us use the notation M_W for the age at death of a white male, and likewise F_W for a white female, M_B for a black male, and F_B for a black female. Then in probability notation

$$P\{55 \le M_W\} = .824,$$

read "the probability that M_W is equal or greater than 55 equals .824". Probabilities are often easier to think of in terms of 1000, or some other convenient large number of cases.

Exercises for Example 1

1.   Calculate

a) $P\{55 \le F_W\}$

b) $P\{65 \le F_B\} - P\{65 \le M_B\}$

c) $P\{1 \le M_W\} - P\{1 \le M_B\}$.

2.   What conclusions would you draw from the numbers in Exercise 1?

EXAMPLE 2

What is the probability that a newborn black male baby will die between ages 45 and 60?

Solution.   .812 - .626 = .186.   That is, out of 1000 black male babies, 812 live to age 45 and 626 of these go on to live to age 60, leaving 812 - 626 = 186 out of 1000, on the average, who die between 45 and 60 (45 through 59).   In probability notation this is expressed as

$$P\{45 \le M_B < 60\} = .186.$$

Exercise for Example 2

1.   Calculate the following probabilities:

   a) $P\{20 \le M_W < 80\}$

   b) $P\{F_B < 25\}$

   c) $P\{F_W < 15 \text{ or } F_W \ge 85\}$.

EXAMPLE 3

What is the average lifetime of a white male?

Solution.   We do not have enough information in the table to answer this question exactly.   However, we can get a lower bound for the average white male lifetime, called $E(M_W)$ in the usual notation, in a very simple way: We will obviously decrease our estimate of $E(M_W)$ below the actual value if we interpret $85 \le M_W$ as $85 = M_W$, that is, if we assume that a man who lives to age 85 does not live past age 85.   Likewise, we will decrease our estimate of $E(M_W)$ if we interpret $80 \le M_W < 85$ as $80 = M_W$, that is, by assuming that a man who dies between ages 80 and 85 actually dies at age 80.

The definition of the mean, it will be recalled, is the weighted sum of the ages.   Each age of death is weighted by the probability that it occurs.   The idea is that we want to add up all the years that are lived by, say, 1000 men, and then divide by 1000.   This is equivalent to weighting by the probabilities.   A probability can be regarded here as the number out of 1000 who died at the stated age.   (See reference [1].)

Proceeding to make the calculation, we need to find $P\{85 \le M_W\}$, $P\{80 \le M < 85\}$, etc., and the table gives $P\{85 \le M_W\} = 0.131$, $P\{80 \le M_W < 85\} = 0.261 - 0.131 = 0.130$, and so on.   Thus

$$E(M_W) > 85(0.131) + 80(0.130) + \ldots,$$

or writing it out as a formula

$$E(M_W) > 85 \cdot P\{85 \leq M_W\} + 80 \cdot P\{80 \leq M < 85\}$$
$$+ 75 \cdot P\{75 \leq M_W < 80\} \ldots + 5 \cdot P\{5 \leq M_W < 10\}$$
$$+ 1 \cdot P\{1 \leq M_W < 5\} + 0 \cdot P\{0 \leq M_W < 1\}.$$

The sum on the right is evaluated from the table as 64.8 years.

We can get a better estimate of $E(M_W)$ if we make more reasonable assumptions about M_W. For example, in the case $80 \leq M_W < 85$ we can "split the difference" and assume $M_W = 82.5$. The only case we cannot easily handle in this way is $85 \leq M_W$, which we will continue to assume means $85 = M_W$. Splitting all the differences in this way gives the approximation

$$E(M_W) = 66.9.$$

(The actual figure, calculated from a more complete table, is 67.6.)

Exercises for Example 3

1.   Calculate approximations for $E(M_B)$, $E(F_W)$, and $E(F_B)$.

2.   It is not surprising that our approximate answer $E(M_W) = 66.9$ is less than the real answer 67.6 since we are still assuming $85 \leq M_W$ means $85 = M_W$.

     However, it is not easy to see whether "splitting the difference" adds too much or too little to the estimate of $E(M_W)$. One way to investigate this is to pretend that our table is missing one column, say the one for age 75, and see how this affects the estimate of $E(M_W)$ obtained by the method above. Do this and see if you then want to change the estimate of $E(M_W)$.

EXAMPLE 4

A black male baby and a white male baby are born on the same day. What is the probability that the white outlives the black?

Solution. Once again we must use approximations. The probability we are interested in is $P\{M_B < M_W\}$, but we cannot directly evaluate this from the table. Let us evaluate a simpler probability first,

$$P\{60 \leq M_B < 65 \text{ and } 65 \leq M_W\},$$

which is the probability that the black male dies between ages 60 and 65 while the white male lives to 65. This probability equals

$$P\{60 \leq M_B < 65\} \cdot P\{65 \leq M_W\} = .099 \times .658 = .065.$$

Here we are making the independence assumption that we can multiply the two probabilities together to get the probability of both of them happening. This is the same assumption that we make when we say "the probability of flipping heads with a fair coin is 1/2, therefore the probability of flipping heads twice in a row is $\frac{1}{2} \cdot \frac{1}{2} = \frac{1}{4}$".

Now notice that $60 \leq M_B < 65$ and $65 \leq M_W$ guarantees that $M_B < M_W$. There are several other similar cases whose probabilities we can calculate from the table by the independence assumption: $0 \leq M_B < 1$ and $1 \leq M_W$, $1 \leq M_B < 5$ and $5 \leq M_W$, etc. If we draw a Cartesian coordinate system with M_B and M_W as the coordinates, we see that these cases divide the region $M_B < M_W$, which is those points above the diagonal line, into strips. For example, $40 \leq M_B < 45$ and $45 \leq M_W$ is the strip marked * in Fig. 1 (page 50).

The total probability in all these strips is

$$P\{0 \leq M_B < 1\} \cdot P\{1 \leq M_W\} + P\{1 \leq M_B < 5\} \cdot P\{5 \leq M_W\}$$

$$+ \ldots + P\{75 \leq M_B < 80\} \cdot P\{80 \leq M_W\}$$

$$+ P\{80 \leq M_B < 85\} \cdot P\{85 \leq M_W\}$$

which is calculated to be .542.

We can see that $P\{M_B < M_W\}$ must be greater than .542 since we still have not included the cases in the heavy squares along the diagonal. For example, the square marked ** in the case $80 \leq M_B < 85$ and $80 \leq M_W < 85$, which has probability

$$P\{80 \leq M_B < 85\} \cdot P\{80 \leq M_W < 85\} = .104 \times .130 = .0135.$$

We do not know how much of this probability belongs to

the event $M_B < M_W$, but once again we can split the
difference and assume that 1/2 of this amount does.
That is, we can add $(.0135)/2 = .0068$ to our estimate
of $P\{M_B < M_W\}$. Essentially we are saying that if the
black man and the white man die in the same 5-year interval, there is probability 1/2 that the black dies
first (probably an underestimate).

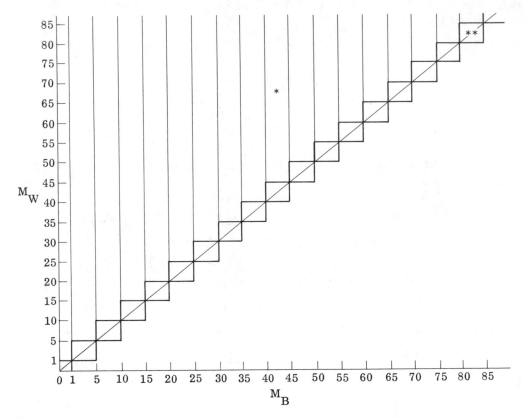

Figure 1

The total probability of the squares along the
diagonal is

$$P\{0 \le M_B < 1\} \cdot P\{0 \le M_W < 1\} + P\{1 \le M_B < 5\} \cdot P\{1 \le M_W <$$

$$+ \ldots + P\{80 \le M_B < 85\} \cdot P\{80 \le M_W < 85\}$$

$$+ P\{85 \le M_B\} \cdot P\{85 \le M_W\}$$

which equals .092. Adding 1/2 of this to our estimate
of $P\{M_B < M_W\}$ gives the approximation

$$P\{M_B < M_W\} = .542 + \frac{1}{2}(.092) = .588.$$

Exercises for Example 4

1.   Accepting this estimate of $P\{M_B < M_W\}$ as accurate,

it must be true that $P\{M_W < M_B\}$, the probability that the black outlives the white, equals $1 - .588 = .412$. Calculate this probability directly in a manner similar to that used above.

2. Calculate $P\{M_W < F_W\}$ and $P\{M_B < F_W\}$.

3. Investigate how splitting the difference affected the estimate of $P\{M_B < M_W\}$, by recalculating it pretending that the table does not go beyond age 80.

EXAMPLE 5

A white male baby and a black female baby are born on the same day and both of them are alive on their 45th birthday. Which has the greater probability of surviving to age 65?

Solution. We are interested in the "conditional probabilities" of living to age 65, given that one has lived to 45. We denote this as $P\{65 \leq M_W | 45 \leq M_W\}$ and $P\{65 \leq F_B | 45 \leq F_B\}$. Thinking of 1000 cases is helpful here: of 1000 white male babies, 905 will survive to age 45 and 658 of these will survive to 65. Therefore, the proportion of those alive at age 45 who survive to 65 is $658/905 = .727$, and this is the value of $P\{65 \leq M_W | 45 \leq M_W\}$. Likewise

$$P\{65 \leq F_B | 45 \leq F_B\} = 641/864 = .742.$$

We see the black female has greater conditional probability of surviving to 65, given that she has lived to 45, even though she has smaller probability than the white male of living to 65. An explanation of this is that women in general tend to outlive men, but in the United States blacks run higher risks of dying young. Given that the black woman has survived this high-risk period, she then tends to outlive the white male.

Exercises for Example 5

1. Calculate $P\{25 \leq F_B | 1 \leq F_B\}$, $P\{25 \leq M_B | 1 \leq M_B\}$, $P\{25 \leq M_W | 1 \leq M_W\}$, $P\{25 \leq F_W | 1 \leq F_W\}$.

2. Recalculate the table for ages 70, 75, 80, and 85 conditional on being alive at age 65.

3. A man aged 40 and a woman aged 35, both white, marry "till death do us part". What is the

probability that she will be a widow?   [Hint: This is the conditional probability

$$P\{M_W < F_W + 5 \mid 40 \leq M_W \text{ and } 35 \leq F_W\}.]$$

Reference

[1]   SBE, Exploring Data, Set 6.

Introduction to the Chi-Square Procedure

ROGER CARLSON

INTRODUCTION

Probabilistic laws are used to predict the outcome of certain experiments. But what if the observed results of the experiment do not agree with the predictions? Of course probabilistic laws do not predict exact results. What we need is some method of checking observations against predictions to see if our observations are sufficiently close to the predictions to make us reasonably confident that the laws actually govern the situation being tested in the experiment. In this paper the chi-square procedure for such checking is presented and illustrated in several different ways. No proof is offered but enough examples are worked out to enable a student to apply the chi-square test to the problems in other sets.

Broadly speaking, the laws of science can be divided into two types: exact laws and probabilistic laws. As an example of the first type, Kepler's laws can be used to predict the motions of the planets around the sun. If we predict the position of Mars for 9:00 P.M. tomorrow night, we expect it to be there then. And if it is not there, we could not include Kepler's laws among the laws of science.
    Mendel's laws of inheritance are of the second general type. They cannot be used to predict the exact kind of offspring from two parents, but they can be used to predict the probability of a certain kind of offspring.

The author is at the University of Missouri at Kansas City, Kansas City, Missouri.

The difference between these types is especially important when we try to test a law by checking it against a set of observations. As we have said, if Mars is not where it is predicted to be, then we can reject the laws of planetary motion. (Unless, of course, there is an error in our calculation of its position or some catastrophe has destroyed the planet.) But Mendel's laws are not so easily rejected. For example, in one experiment he predicted that a certain pea plant would have as offspring yellow peas with probability 3/4. Out of 36 peas, 25 were yellow and the law was accepted, even though the number of peas predicted was 27, that is, $27 = \frac{3}{4}(36)$. We must recognize that for a probabilistic law deviations from the law can occur by "chance" and statistical procedures have to be developed to judge the agreement of observation with the theory.

In this example we discuss one method for judging the agreement of observation with a probability law. This is called the chi-square procedure and was first introduced by Karl Pearson in 1900.

We begin by learning how to calculate χ^2 which is used to measure the closeness of observed values to a set of predicted values. Since we are dealing with probability predictions and not exact laws, the value of χ^2 can vary by "chance". Therefore we need to look at the probability distribution of χ^2 in order to apply it as a test of a theory. We then present several examples of the use of the chi-square procedure. The letter χ is the lower-case greek letter chi, pronounced "ky", rhyming with "sky".

The method of calculating chi-square can be illustrated by the following example.

CALCULATION OF χ^2

Two coins are tossed 100 times and we record the number of times that no heads, one head, and two heads come up. These numbers are 20, 54, and 26, respectively. We want to test the "theory" that the coins are both fair (and fairly tossed) and that the outcome of one of the coins in no way influences the outcomes of the other coin. From this theory we predict that in 100 tosses on the average, the coins will fall with no heads up about 25 times; one head up about 50 times, and both heads up about 25 times. These numbers 25, 50, and 25, which are predicted, are called the expected numbers for the outcomes. They will be the same for every set of 100 tosses and are to be compared with the observed outcomes which for this particular experiment happen to be 20, 54, and 26. Table 1 shows the results. (The expected numbers were obtained by first calculating the probability of 0, 1, and 2 heads and multiplying each of these probabilities by 100. For example, the probability of 0 heads is (1/2)(1/2) = 1/4 and so the expected number of heads is 100(1/4) = 25.)

Table 1. Observed and expected number of heads for 100 tosses of two coins

Number of heads	Expected number	Observed number
0	25	20
1	50	54
2	25	26
Total	100	100

We now need to make a comparison of the numbers 25, 50, and 25 with the numbers 20, 54, and 26 and, as the statistician would say, test the "goodness of fit" of these data with our theory. There are many ways of making such a comparison and we do not claim that this one is necessarily best. It is however a reasonable method and simple to calculate, and there are extensive tables available for its use.

Here we will measure the agreement of theory and observation by calculating:

$$\chi^2 = (20-25)^2/25 + (54-50)^2/50 + (26-25)^2/25$$

$$= 34/25 = 1.36.$$

(The symbol is chi-square and it is called the chi-square statistic because the Greek letter chi, χ, is traditionally used when discussing its probability distribution.)

To explain exactly why χ^2, computed as above, is a reasonable way to measure the agreement of theory and observation would require a long inquiry into statistical theory. All we can do here is point out a few things which might help explain why χ^2 is used.

To calculate χ^2 from Table 1 we square the difference between the observed value and the expected value for each row, divide each square by its expected value, and then add all these. Why? First, the difference of the observed and expected value is certainly what we are interested in, and if the observations agree exactly, $\chi^2 = 0$. But if we just add these differences we get 0. That is, (20-25) + (54-50) + (26-25) = 0. More generally, for this experiment the observed numbers always add to 100 and the expected values always add to 100 so the sum of their differences is 0. Of course, we didn't have to square them to avoid this problem, but squares are fairly easy to work with and sums of squares occur in many other branches of statistics so that chi-square can be easily related to these other theories.

The hard thing to explain about χ^2, at this point, is the division of each square by its expected number. Basically, the reason for this is that the variation of the observed numbers for each row depends on the number we expect for that row. The expected number, in turn, depends on the number of trials and the probability for that row so that we are "adjusting" by this division for sample size and putting all cells on an equal footing with regard to the probabilities. This is analogous to looking at the percentage deviations rather than the deviations themselves. As before, this does not explain why we divided by the expected number. Why not divide each square by the square of the expected number? The answer, as before, is that dividing by the expected number rather than the square of the expected number simplifies the theory of the chi-square procedure. We have thus made the ideas in the formula plausible, but we have not given a defense for the specific functional form used, except convenience.

So much for justifications. The calculation of χ^2 in general is given by

$$\chi^2 = (O_1 - E_1)^2/E_1 + (O_2 - E_2)^2/E_2 + \ldots + (O_k - E_k)^2/E_k$$

where $O_1, O_2, \ldots, O_k$ are the observed numbers for each cell and $E_1, E_2, \ldots, E_k$ are the expected numbers for each cell. (E_1 for example is the number of trials times the probability for the first cell.)

Exercise

1. Table 2 shows the number of times 0, 1, and 2 heads were obtained in four sets of 20 tosses of two coins. For the first set

   $$\chi^2 = (4-5)^2/5 + (11-10)^2/10 + (5-5)^2/5$$

   $$= 3/10 = 0.30.$$

   Calculate χ^2 for each of the other three sets. Which set has the largest value of χ^2? Which set has the smallest value of χ^2? Why? Collect three sets of your own by tossing two coins 60 times. Calculate χ^2 for each set. Did you get the same value of χ^2 for all three sets?

Table 2. Observed number of heads for four
sets of 20 tosses of two coins

Number of heads	Expected (for fair coin)	Set 1	Set 2	Set 3	Set 4
0	5	4	8	8	9
1	10	11	8	6	1
2	5	5	4	6	10

EXAMPLE 1

Table 3 gives the outcomes of 240 tosses of a fair die.
These were grouped into four sets of 60 tosses each.
For each set we calculated χ^2 based on the prediction
that each of the faces was equally probable so that for
60 tosses each face was predicted to appear 10 times.
The values of χ^2 are shown in the last column of Table
3. Notice that they are not the same. That is, the
value of χ^2 varies for repetitions of the same experiment.

Table 3. Four sets of 60 tosses of a fair die

Number of spots	1	2	3	4	5	6	Total	χ^2
Predicted	10	10	10	10	10	10	60	
Set 1	7	12	8	14	10	9	60	3.4
Set 2	12	11	10	8	10	9	60	1.0
Set 3	13	10	11	9	9	8	60	1.6
Set 4	11	10	10	12	9	8	60	1.0

THE CHI-SQUARE PROBABILITY DISTRIBUTION

We have now learned how to calculate χ^2 for a set of
observed and expected values. We have also seen some
reasons why statisticians would use χ^2 as a measure of
"goodness of fit". Thus in Exercise 1 we saw that "bad"
fits like set 4 give larger values of χ^2 than "good"
fits like set 1. But in Example 1 we saw that χ^2 will
vary over repetitions of the same experiment. If a die
is tossed 60 times we get one value for χ^2, but for 60
more tosses of the same die we may get another value
for χ^2. Thus in order to use χ^2 to compare a particular set of data to a theory, we need to know how much
variation χ^2 can have over repetitions of the same
experiment, when the theory is true. In other words we
need to know the probability distribution of χ^2 to tell

whether the values of χ^2 we have are larger than can be expected by chance. Table 4 is a short table of these probabilities. It is a table of the <u>chi-square</u> (χ^2) probability distribution. Before going on we should see how to read this table.

Table 4. Table of chi-square distribution

Degrees of freedom	Probability			
	.20	.10	.05	.01
1	1.64	2.71	3.84	6.63
2	3.22	4.60	5.99	9.21
3	4.64	6.25	7.81	11.34
4	5.99	7.78	9.49	13.28
5	7.29	9.24	11.07	15.09
6	8.56	10.64	12.59	16.81
7	9.80	12.02	14.07	18.47
8	11.03	13.36	15.51	20.09
9	12.24	14.68	16.92	21.67
10	13.44	15.99	18.31	23.21
15	19.31	22.31	25.00	30.58
20	25.04	28.41	31.41	37.57
25	30.67	34.38	37.65	44.31
30	36.25	40.26	43.77	50.89

Exercise

2. Table 4 has 14 rows and 4 columns. Each row of the table corresponds to a different distribution of chi-square depending on its <u>degrees of freedom</u> (a term which we will explain later) and we have given values for 1, 2, ..., 10, 15, 20, 25, and 30 degrees of freedom. The <u>columns</u> of the table refer to probability levels. Thus in row 5, column 2 of the table we find the number 9.24. This means that if we are comparing observations with a theory using 5 degrees of freedom, then the probability that χ^2 will be <u>larger</u> than 9.24 is .10 if the theory is true. In other words, chi-square can vary over repetitions of the same experiment but 90% of the time it will be less than 9.24 (with 5 degree of freedom). You can see from this table that for 8 degrees of freedom chi-square will be larger than 11.03 20% of the time, that it will be larger than 13.36 10% of the time, and so on. Familiarize yourself with the table by finding the following probabilities for chi-square.

a) $P(\chi^2 > 12.02)$ for 7 degrees of freedom.

b) $P(\chi^2 > 50.89)$ for 30 degrees of freedom.

c) $P(\chi^2 > 1.64)$ for 1 degree of freedom.

Now use the table to solve for x in (d), (e), and (f) below.

d) $P(\chi^2 > x) = .10$ for 10 degrees of freedom.

e) $P(\chi^2 > x) = .01$ for 3 degrees of freedom.

f) $P(\chi^2 > x) = .20$ for 30 degrees of freedom.

THE DEGREES OF FREEDOM FOR CHI-SQUARE

The probability distribution of chi-square depends only on the number of cells in which the observations are grouped. The rule is simple: If we arrange our observations in k cells for testing goodness of fit the probabilities for χ^2 are given approximately by the chi-square distribution with k-1 degrees of freedom. Thus to know which row of the table to use we merely count the number of cells for the data and subtract 1. For the data of Table 3 (60 tosses of a fair die) there are 6 cells and so we have 5 degrees of freedom. From Table 4 we see that about 80 per cent of the time χ^2 will be less than 7.29; the largest value of χ^2 is 3.4 and we would have to agree that these are indeed tosses of a fair die.

EXAMPLE 2

Statisticians often use tables of random digits (see reference [1].) and before such tables were readily available they would sometimes use other sources of digits--such as telephone directories--in the hope that these would provide random digits. The chi-square procedure gives us a method of testing whether such sources give truly random digits. Table 5 shows the results of such a test. Ten thousand digits were taken from a telephone directory by selecting pages haphazardly, ignoring numbers printed in heavy type and taking the last two digits of the remaining numbers. The frequency of the digits 0, 1, ..., 9 are shown. The probability for any particular digit is 1/10 and so we would predict that in 10,000 trials each digit would appear 1,000 times. Of course we wouldn't get exactly 1,000 of each; the reason for the chi-square test is to see how much variation from the predicted numbers can be allowed. As you see χ^2 here is 58.55. There are 10 cells in Table 5. Since the numbers must add to 10,000, the first 9 numbers enable us to compute the 10th one. Therefore chi-square has 9 degrees of freedom. Now look at Table 4 for 9 degrees of freedom. The numbers increase as the probabilities decrease and our 58.55 is way out at the end of the distribution. The probability that chi-square is greater than 58.55 is therefore much smaller than .01 and we must seriously question the randomness of these digits.

Table 5. χ^2 for digits from a telephone directory

Digit	Observed frequency	Expected frequency	O-E	$(O-E)^2$	$(O-E)^2/E$
0	1026	1000	26	676	.68
1	1107	1000	107	11449	11.45
2	997	1000	-3	9	.01
3	966	1000	-34	1156	1.16
4	1075	1000	75	5625	5.62
5	933	1000	-67	4489	4.49
6	1107	1000	107	11449	11.45
7	972	1000	-28	784	.78
8	964	1000	-36	1296	1.30
9	853	1000	-147	21609	21.61

$$\chi^2 = 58.55$$

Since we have reason to believe that the telephone directory (or at least this one) is a poor source of random digits, we might ask why. We should look at Column 4 of Table 5 to see which digits occur more or less frequently than we predicted. The "offenders" are the digits 1, 6, and 9, with 9 occurring less often than we expect and 1 and 6 occurring more often. Maybe the reader can suggest why these deviations would occur in a telephone directory.

ANOTHER USE OF CHI-SQUARE

The basic ideas of chi-square--that of measuring "goodness of fit" by using χ^2--can be applied to other statistical problems. The two examples below show how χ^2 can be used to explore the relationship between two categories of a 2 × 2 table (read: "2 by 2 table").

EXAMPLE 2

Two coins (a penny and a nickel) were tossed 100 times and the results arranged in the following table:

		Penny Head	Penny Tail	Total
Nickel	Head	20	31	51
	Tail	23	26	49
	Total	43	57	100

Here we are interested in whether the outcomes on the penny influence the outcomes for the nickel. (In statistical language, we are testing the independence of the tosses for the penny and the nickel.) Both coins fell heads in 20 of our 100 tosses. How many head-head outcomes would we predict if the two coins were independent? To calculate this notice that the penny fell heads in 43% of its tosses, while the nickel fell heads in 51% of its tosses. If the coins do not influence each other we would expect that in 43% of the tosses where the nickel fell heads the penny would also fall heads. That is, we would expect the head-head outcome in (.43)51 = 22 tosses. As before we must allow some variation from the predicted value of 22 and we will use χ^2 to measure the amount of allowable variation.

First we must complete the table of predicted outcomes. To do this we make another 2 × 2 table containing the totals for both coins and enter in this table the predicted value of 22 for head-head. This is the table at the left below. We see that the rest of the table can be filled in from these numbers, giving the table on the right. (For example, if we predict 22 head-head outcomes then we must also predict that the penny will fall tails and the nickel heads in 29 tosses because 22 + 29 must add to 51.)

		Penny Head	Tail	Total			Penny Head	Tail	Total
ickel	Head	22		51	Nickel	Head	22	29	51
	Tail			49		Tail	21	28	49
	Total	43	57	100		Total	43	57	100

Since that table could be completed by knowing the outcome for only one cell (we used the head-head cell), the degrees of freedom for chi-square is 1. As before, χ^2 is computed from the observed and expected values:

$$\chi^2 = (20-22)^2/22 + (31-29)^2/29 + (23-21)^2/21$$

$$+ (26-28)^2/28$$

$$= .18 + .14 + .19 + .14 = .65.$$

To see if χ^2 is too large for independent coins we use row 1 of Table 4. There we see that with 1 degree of freedom chi-square will be less than 1.64 in about 80% of all such experiments and therefore our value of .65 is not exceptionally large. We conclude that the two coins do not tend to influence each other.

(Notes on Example 2. (1) Forty-three percent of 51 is 21.93, not 22. We used 22 to simplify our discussion. The reader should recalculate χ^2 using 21.93 and adjusting the other predicted values accordingly, to see if this round off makes a difference in our conclusion. (2) We had 1 degree of freedom here because we could complete the table of predicted values from one of the cells. This is a general rule: The number of degrees of freedom for a table with r rows and c columns is the minimum number of cells needed to complete the table of predicted values.)

EXAMPLE 3

A poll is conducted to investigate the readership of two magazines. Call them magazines A and B. One hundred people were interviewed with the following results:

	Read A	Do not read A	Total
Read B	16	14	30
Do not read B	24	46	70
Total	40	60	100

For example, 30 people said that they read B and 16 of these said that they also read A. Let's see if there is some connection between the two magazines in the sense that people tend to read both if they read one of them. To represent "no connection", we will think of the results as if each person tossed two (biased) coins in order to decide on reading the magazines. If the penny and nickel both fall heads, he reads both A and B. If the penny falls heads and the nickel falls tails, then he will read A but not B. And so on. What results would we predict if the two magazines are unrelated? Since 30 people read B and the overall percentage of A readers is 40%, we would expect that 40% of the 30 would read both A and B; (.40)30 = 12. Thus there are 12 people in the A-and-B cell of the predicted table and the other numbers can be filled in from the totals. This gives:

	Read A	Do not read A	Total
Read B	12	18	30
Do not read B	28	42	70
Total	40	60	100

As before

$$\chi^2 = (16-12)^2/12 + (14-18)^2/18 + (24-28)^2/28$$

$$+ (46-42)^2/42$$

$$= 16/12 + 16/18 + 16/28 + 16/42 = 3.17.$$

From the chi-square table for 1 degree of freedom we see that 10% of the time χ^2 would be bigger than 2.71 and that 5% of the time it would be less than 3.84. Our value of 3.17 is in between these two and we do not have a clear-cut case. Slightly more people tend to read both magazines than we predicted (16 as opposed to 12 predicted) but the overall deviation is not overwhelming. Some personal judgment would have to go into saying whether A and B were related.

SUMMARY

The formula for χ^2 is

$$\chi^2 = (O_1-E_1)^2/E_1 + (O_2-E_2)^2/E_2 + \ldots + (O_k-E_k)^2/E_k$$

where $O_1, O_2, \ldots, O_k$ are the observed occurrences and $E_1, E_2, \ldots, E_k$ are the predicted occurrences of k events.

It can be used as a measure of agreement of observations with a theory, although it is not the only such measure available. Historically it is the first one to be proposed. Large values of χ^2 indicate that the observations do not fit the theory; for perfect agreement $\chi^2 = 0$. To know how large χ^2 should be in order to challenge a particular set of data we need to use the table of the chi-square probability distribution. This tells us the approximate probability that χ^2 will be larger than a particular number and our judgment of the goodness of fit is then based on this probability. The chi-square table used depends on the number of degrees of freedom which is (a) the number of cells minus 1 for a table of k cells and (b) 1 for a 2 × 2 table.

Exercises

1. Toss a penny and a nickel 100 times and record the number of times the four possible outcomes occurred. Arrange your results in a 2 × 2 table and apply the test of Example 2. Now regroup the results according to the number of heads appearing on the two coins. Calculate χ^2 for this arrangement to see if both coins are "fair".

2. The table below shows results reported by Gregor Mendel in one of his experiments with pea plants. Each pea can be round or wrinkled and it can be

yellow or green, thus giving four possible cells. According to Mendel's theory the proportions of the four outcomes is 9:3:3:1, giving the expected numbers shown in the table. (These were rounded into integers to simplify the arithmetic.) Look at the table to see if you would say that the data confirms the theory. Now calculate χ^2 and use the table for 3 degrees of freedom to test the agreement.

Pea type	Observed number of peas	Expected number of peas
Round and yellow	315	313
Wrinkled and yellow	101	104
Round and green	108	104
Wrinkled and green	32	35
Total	556	556

3. A remedy is proposed for the common cold and a doctor is skeptical of its value as a cure. He decides to test the new remedy by giving it to some of his patients while treating another group of patients by the usual method of treatment. To get a fair comparison of patients he tosses a coin for each cold sufferer and gives him the new remedy if the coin falls heads and gives the standard treatment to those patients for whom the coin falls tails. (He knows the new remedy will do no harm and is at least as good as the old remedy so that the choice by coin-toss is not a mistreatment of any patient.) The results are shown below. Calculate χ^2 as in Example 2 and say what advice you would give to the doctor for treating cold patients in the future. Suppose that the new treatment was very expensive relative to the older method of treating colds; would this affect your advice?

	Relieved	Not relieved	Total
New treatment	31	15	46
Old treatment	29	25	54
Total	60	40	100

4. We are told that a coin fell heads 51% of the time. Is is a fair coin? Make three tables of

observed and expected values each having 51%
observed heads but the first table is for 100
tosses, the second table for 1000 tosses and the
third table is for 10,000 tosses. Calculate χ^2
for each table. Explain why just giving the percentage of heads is not enough to judge the fairness of a coin.

Reference

[1] SBE, Weighing Chances, Set 2.

Stock Market Fractions

FREDERICK MOSTELLER

INTRODUCTION

We use the χ^2 test to check whether the distribution of final stock market fractions is uniform.

EXAMPLE

A table of the number of eighths in the fraction on a closing stock price for low-priced stocks from the American Stock Exchange gave the frequencies in Table 1. (The final stock fractions were discussed in reference [1].)

Table 1. Frequency distribution of final fraction on stocks selling at or below 10 on the American Stock Exchange

Fraction	Eighths	Frequency
0	0	60
1/8	1	30
1/4	2	29
3/8	3	27
1/2	4	47
5/8	5	49
3/4	6	37
7/8	7	38
		317

The author is at Harvard University, Cambridge Massachusetts.

Find out if the data in Table 1 are compatible with equally likely eighths by using the chi-square statistic.

<u>Solution</u>.   In the special case where all expected values are the same, the formula for chi-square simplifies considerably.   For each count we need (observed count - expected count)^2/expected count, and then this quantity must be summed over the counts.   In our problem the observed counts are $O_1, O_2, \ldots, O_8$ and the expected counts are all equal in the simple model we have chosen.   Let us call the expected count for a cell E.   We need to compute

$$\frac{(O_1 - E)^2}{E} + \frac{(O_2 - E)^2}{E} + \ldots + \frac{(O_8 - E)^2}{E} \; .$$

The first term is

$$(O_1^2 - 2O_1 E + E^2)/E$$

and the others are similar.   When we add them all up, the lead terms give

$$\frac{O_1^2 + O_2^2 + \ldots + O_8^2}{E} \; .$$

The sum of the O's is the sample size N, and so the sum of the middle terms gives $-2N$, and the sum of all the E's is also N.   All told we get

$$\chi^2 = \frac{\text{Sum of the } O^2\text{'s}}{E} - N.$$

We need the sum of the squares of the O's for our problem:

O_i	O_i^2
60	3600
30	900
29	841
27	729
47	2209
49	2401
37	1369
38	1444

$$\Sigma O_i^2 = 13493$$

$$\chi^2 = \frac{13493}{317/8} - 317$$

Approximately this gives 340 - 317 = 23.

This chi-square has 7 degrees of freedom and we find that a value of .23 is far outside Table 4 of Set 6 in this book. This is a huge departure from the uniform model. We conclude that even for the low-priced stocks, the eighths are not equally frequent. The largest contribution comes from zero eighths.

Exercises

1. Show that 0 contributes about 10 to χ^2 in Table 1.

2. Delete the cases for 0 and see if it is reasonable that the seven eighths that are left are equally distributed, using the chi-square formula. Use 6 degrees of freedom.

3. Write out and justify in detail the steps in the derivation of the simplified formula for χ^2 when the expected values are all equal.

Reference

[1] SBE, Exploring Data, Set 2.

Independence of Amoebas

JOEL E. COHEN

INTRODUCTION

What effects are caused by the presence of two different kinds of amoebas in the intestinal tracts of men?
One kind may cause disease and the other not, but the
presence of the second kind may also prevent the
disease usually caused by the first. A simple 2 × 2
table yields a chi-square test that helps to decide
whether infection by one kind leads to infection by the
other.

An epidemic of severe intestinal disease occurred
among the workers in a woodworking plant in South Bend,
Indiana. Doctors attributed the illnesses to unusually large numbers of an amoeba, called Entamoeba
histolytica, growing in the men. After investigating
alternative possibilities, public health officials concluded that the amoebae entered the plant's water
supply through a leaky water main which passed near a
leaky sewage pipe.

In order to see how many of the men in the plant
who had not fallen sick had been infected with the
amoeba unawares, the public health officials chose a
random sample of 138 apparently well workers in the
plant. The results are summarized in Table 1. They
found the amoeba in the intestinal contents of 70 of
the men. They distinguished two races of Entamoeba
histolytica, large and small. (In the large race, the
average diameter at one stage in the life cycle is

The author is at Harvard University, Cambridge,
Massachusetts.

greater than 10 microns, while in the small race, the average diameter at the same stage is less than 10 microns.) They found the large race only in 35 men, the small race only in 23 men, and both races in 12 men.

Table 1. Observed frequencies of infections with the large and small races of Entamoeba histolytica among 138 randomly chosen workers in a South Bend (Indiana) woodworking plant

	Large race		
Small race	Present	Absent	Totals
Present	12	23	35
Absent	35	68	103
Totals	47	91	138

Since the small race of the amoeba is believed not to cause disease, we can ask some interesting questions of their data. Does infection with one race confer resistance to infection with the other? Or does infection with one race increase susceptibility to the other? In short, do the two races interact in infecting man?

We recall that if two events A and B are independent, then the probability that both occur is the product $P(A)P(B)$. If we have several trials, n, then the expected frequency with which both occur is $nP(A)P(B)$. Let $\bar{A}$ be the event complementary to A and $\bar{B}$ the event complementary to B.

In a 2 × 2 table, if the events A and B are independent, then so are the pairs of events $\bar{A}$ and B, A and $\bar{B}$, and $\bar{A}$ and $\bar{B}$, as you are asked to prove in Exercise 6. And so we find if A and B are independent, the expected counts in a 2 × 2 table are

	B	$\bar{B}$	Total
A	$nP(A)P(B)$	$nP(A)P(\bar{B})$	$nP(A)$
$\bar{A}$	$nP(\bar{A})P(B)$	$nP(\bar{A})P(\bar{B})$	$nP(\bar{A})$
Total	$nP(B)$	$nP(\bar{B})$	n

Note that the upper left-hand corner cell

$$nP(A)P(B) = \frac{(\text{row total})(\text{column total})}{n}$$

and that the formula on the right can be used to get every cell entry in the 2 × 2 table.

In the numerical tables the row and column totals are estimates of nP(A) and the other marginal totals.

Let us return to the data of Table 1. If the presence or absence of the large race made no difference to the frequency with which the small race was present or absent, then among the total 47 individuals in which the large race was found, the small race should be present in the fraction 35/138, that is, in 47 × 35/138 = 11.9 individuals. The expected numbers of individuals with or without each kind of infection, assuming no interactions between the infections, are presented in Table 2.

Table 2. Predicted frequencies of infections with the large and small races of Entamoeba histolytica among 138 randomly chosen workers, assuming the same total prevalence of each race and no interaction between the races

Small race	Large race		
	Present	Absent	Totals
Present	11.9	23.1	35.0
Absent	35.1	67.9	103.0
Totals	47.0	91.0	138.0

The observed frequencies and those expected assuming no interaction are so close that the data provide no grounds for supposing that infection with one race of the amoeba makes an individual more or less susceptible to infection with another.

Exercises

1. Show arithmetically that if the expected count for the cell with 23 men is also made from the formula (row total)(column total)/(grand total), the expected values in the row must add to the row total exactly.

2. Generalize the result of Exercise 1 to the table with original cell entries

a	b
c	d

3. In the example, given the row and column totals,

what cell entries, if any, would have come closer to the expected values than the ones found?

4. What would be the expected value of chi-square for a 2 × 2 table if the rows and columns were independent?

5. Computing chi-square for the present numerical example does not seem worthwhile. Why not?

6. Prove that if A and B are independent, so are $\overline{A}$ and B.

7. Explain the practical importance of the discovery of independence between the races of amoeba. Explain what would have been the importance of finding a strong dependence.

Reference

[1]   Charles A. LeMaistre et al., "Studies of a water-borne outbreak of amebiasis, South Bend, Indiana", Am. J. Hyg. 64 (1956): 30-45.

What is the Sample Size?

FRANK W. CARLBORG

INTRODUCTION

Before starting a scientific experiment one needs to design it carefully to be sure the results, if any, will answer the question posed. Sometimes what seem like many measurements leading to highly reliable answers turn out to be worth less than a single measurement, as this set illustrates. The author explains what further corrective steps are needed.

THE EXPERIMENT

A developer of an experimental germicidal soap was interested in the soap's ability to prevent secondary infections from sores, cuts, and abrasions. Previous laboratory work had demonstrated the desired effect on the usual bacteria associated with these infections. A small clinical program had been favorable and had suggested efficacy in actual use. The goal, then, was to run a large, conclusive experiment. Getting cooperation from some large, residential institution seemed like the best opportunity for a controlled experiment. Persons in residence would use the soap. They would develop sores, cuts, and abrasions naturally, and the incidence of infection would be noted.

Exercises

1.   Suppose a state-run home for children and a state university are available institutions. Which

The author is a statistical consultant in St. Charles, Illinois.

1

seems like the better site for this experiment?

2. Suppose the experiment is to run for one month. Which seems better: July or January?

3. Assume the site and time are selected. As the statistical consultant, what do you recommend next?

The company obtained the cooperation of a state-run home for children, with several hundred residents. The plan was to have about half the children wash regularly with the germicidal soap and the other half wash with a placebo bar. The placebo was identical to the test bar except that the germicide was not included. Assignment of a soap to a child was made randomly. When a child developed a sore, cut, or abrasion, the healing history of that wound was recorded on a daily basis. Some of the wounds developed secondary infections, and some did not. The nurse who evaluated the wound did not know which soap was used by a child.

The children using the placebo incurred 206 wounds: 40 developed secondary infections and 166 did not. The overall infection rate for the placebo was .1942 = 40/206. The children using the germicidal soap incurred 193 wounds: 69 developed secondary infections and 124 did not. Their overall infection rate was .3575 = 69/193. Thus the children with the germicidal soap developed infections at a considerably higher rate, .3575 to .1942.

This observed difference in infection rates seems not to be readily explained by chance variation when we apply chi-square to the 2 × 2 table:

	Number with infection	Number without infection	Total
Placebo	40	166	206
Germicidal	69	124	193
Total	109	290	399

Under the assumption of no difference in soaps, the overall infection rate is .2732 = 109/399. The corresponding expected frequencies are:

	Number with infection	Number without infection	Total
Placebo	56.3	149.7	206
Germicidal	52.7	140.3	193
Total	109	290	399

The chi-square statistic associated with this difference in rates is

$$12.6 = \frac{15.8^2}{56.3} + \frac{15.8^2}{149.7} + \frac{15.8^2}{52.7} + \frac{15.8^2}{140.3}$$

(with continuity correction). From the chi-square table with one degree of freedom, the critical value is 10.8 for the .001 probability level. Therefore, the observed difference in infection rates seems not readily explained by chance. That is, the germicidal soap appears to have <u>raised</u> the secondary infection rate.

Continuity correction. To make the use of tables of chi-square give better approximations in 2 × 2 tables the absolute value of (observed count - expected count) is reduced by 1/2 before squaring. Thus in the first term of the example, O-E was 40 - 56.3 = -16.3. This is reduced by 1/2 to -15.8. The same correction is made for each cell.

Exercises (continued)

4. Verify the value 56.3 in the table above.

5. Verify the value 15.8^2 in the numerators above.

<u>A re-evaluation</u>. This conclusion contradicted a fair amount of previous experience. Before making a final decision about the test soap, the company sought a fresh, outside statistical evaluation of the experiment and the data.

The reviewers of the actual experimental procedure recognized a previously ignored factor. Each child lived in one of ten different dormitories in the home. The dorms, not the individual children, were randomly assigned to the two soaps, five dorms to each soap. A dorm used only one soap in the experiment. All the children in a dorm used the same soap. This fact <u>in-validates</u> the calculation made above. That testing assumes that individuals were independently assigned to the soaps and takes no proper account of their living in groups with the same soaps.

If there was a dorm-to-dorm variation in infection rates, then this factor must be considered in the

comparison of the two soaps. If a dorm had a naturally high infection rate, then the soap assigned to that
dorm would tend to look bad.

There was, in fact, a considerable dorm-to-dorm
variation in infection rates. To see this, consider
the five dorms on the placebo:

Dorm	Number of infections	Number without infections	Total number	Infection rate
P-1	18	36	54	.3333
P-2	5	35	40	.1250
P-3	11	37	48	.2292
P-4	4	27	31	.1290
P-5	2	31	33	.0606
Total	40	166	206	.1942

Next consider the five dorms on the germicidal soap:

Dorm	Number of infections	Number without infections	Total number	Infection rate
G-1	6	19	25	.2400
G-2	20	27	47	.4255
G-3	11	30	41	.2683
G-4	30	26	56	.5357
G-5	2	22	24	.0833
Total	69	124	193	.3575

To test the significance of the apparent dorm-to-dorm
variations within a soap, assume constant rates, .1942
and .3575. The expected frequencies under this assumption and the differences from the actual frequencies
are in the table on page 79. The chi-square statistic
for assessing the observed variation in rates is

$$32.4 = \frac{7.5^2}{10.5} + \frac{7.5^2}{43.5} + \frac{2.8^2}{7.8} + \dots + \frac{6.6^2}{15.4} \ .$$

From the chi-square table with eight degrees of freedom, the critical value for the .001 probability level
is 26.1. Therefore, the observed dorm-to-dorm variation within a soap is not readily explainable by
chance. This factor must be considered and overcome in
a comparison of the two soaps. Relative to this factor,
the sample size was only ten, the number of dorms.

Placebo soap

	With infections		Without infections	
Dorm	Actual	Expected	Actual	Expected
P-1	18	10.5	36	43.5
P-2	5	7.8	35	32.2
P-3	11	9.3	37	38.7
P-4	4	6.0	27	25.0
P-5	2	6.4	31	26.6

Germicidal Soap

	With infections		Without infections	
Dorm	Actual	Expected	Actual	Expected
G-1	6	8.9	19	16.1
G-2	20	16.8	27	30.2
G-3	11	14.7	30	26.3
G-4	30	20.0	26	36.0
G-5	2	8.6	22	15.4

We do not use the continuity correction in tables with more than one degree of freedom.

Exercises (continued)

6. Calculate the observed chi-square statistic for the placebo dorms only. How many degrees of freedom are there? Is the statistic significant at the .05 level?

What we have found does not change the direction of the data. After all, the dorms were randomly assigned to the soaps. But it does say that the evidence is not so strong since it is not based upon 399 independent people, but upon 10 dormitories. Indeed, further analysis showed that the dorm-to-dorm variation was so strong as to leave the results rather inconclusive. The company may wish to study this matter further if it can get an appraisal of infectiousness of dormitories, or otherwise get around the problem.

Beyond this, it is not unknown for a germicide to be effective against the germs it attacks, but in addition to be effective against the germ's parasites as well. The result can be that a germicide might be effective in the laboratory and harmful in practice.

The company will be well aware of this. This makes investigations like the one described here all the more important. And as you have seen, it is easy to make a mistake in the analysis.

The type of error made in the original analysis here is a very frequent one. It often goes under the name "inflated sample size". To avoid it one has to have a clear idea of what units can be treated as independent.

Ratings of Typewriters

FREDERICK MOSTELLER

INTRODUCTION

Twenty-five typists tested and rated twenty-five type-writers; the 625 results are tabulated. This example discusses methods of studying the table to determine whether some typists were hypercritical and others indifferent. Were some typewriters really better than others, or were they of about the same quality but preferred by a certain group of typists, while others were preferred by another group? What are the probabilities of getting a table similar to this one if the typists simply assigned random preference ratings to the typewriters before even trying them?

RATINGS OF TYPEWRITERS

Twenty-five typists were trained to rate typewriters--they assigned a typewriter a rating of 1, 2, 3, 4, or 5, 5 being very satisfactory, and 1 being very unsatisfactory. Each typist rated the same 25 typewriters, one each of 25 different brands. They used them to type several different kinds of things--letters, envelopes, postcards, tables, and so on. When they were finished, their ratings produced the data shown in Table 1.

The author is at Harvard University, Cambridge, Massachusetts. Data was kindly provided by Consumers Union, publisher of Consumer Reports, and brought to the author's attention by Cuthbert Daniel. Reprinted by permission.

Table 1. Typewriter ratings from a low of 1 to a high of 5 given by 25 trained raters to 25 brands of typewriters

Brand	Rater 1	2	3	4	5	6	7	8	9	10	11	12	13	14	15	16	17	18	19	20	21	22	23	24	25	Total
1	3	3	3	3	3	4	3	2	4	4	3	2	4	3	5	3	5	4	5	4	3	5	2	3	4	87
2	4	5	3	3	4	5	4	4	4	4	5	3	3	5	5	5	3	4	4	4	4	5	4	3	4	100
3	4	3	2	3	3	5	2	3	3	3	2	2	4	2	3	4	5	4	3	4	3	3	3	4	3	81
4	2	2	1	1	2	1	1	2	2	2	1	1	1	2	1	3	2	1	2	2	1	1	1	2	1	39
5	4	4	3	4	3	5	3	4	3	3	4	3	2	5	5	4	5	4	2	5	4	4	3	4	3	94
6	2	2	2	3	2	3	2	3	3	2	3	3	2	3	3	3	4	2	3	2	2	4	2	2	2	63
7	1	1	1	1	1	1	1	1	1	1	1	1	2	1	1	1	2	1	1	1	1	1	1	1	1	26
8	2	2	1	2	3	3	1	1	3	3	3	2	2	4	3	4	2	3	4	2	3	2	2	1	3	62
9	2	2	4	3	3	2	4	4	4	4	4	2	1	5	5	3	3	5	4	3	3	3	1	3	4	86
10	3	5	4	3	3	4	5	4	4	3	5	4	4	4	5	3	4	2	5	5	4	5	4	3	3	96
11	1	4	3	4	1	4	3	3	4	5	5	3	5	1	5	4	4	1	2	4	4	4	3	3	2	86
12	2	3	3	2	3	5	4	4	5	3	3	3	2	2	3	3	4	5	3	3	3	3	3	2	3	75
13	2	2	1	1	2	1	1	1	2	2	1	1	1	1	2	3	1	1	1	1	2	2	2	1	1	38
14	2	2	2	4	3	3	2	3	3	3	3	2	4	4	2	3	4	4	2	4	2	3	2	3	2	66
15	3	5	4	3	3	5	5	4	5	4	5	4	3	5	5	4	4	4	5	5	5	5	4	3	5	108
16	3	4	4	3	3	3	2	1	4	2	2	2	2	4	4	3	2	1	2	4	4	2	3	2	4	72
17	4	3	4	4	3	5	4	4	5	4	3	3	5	5	4	4	4	5	5	5	3	4	3	3	3	101
18	1	2	2	1	2	1	1	1	1	3	2	2	2	3	2	3	3	1	3	2	1	1	2	2	1	49
19	3	3	3	3	3	4	3	3	3	3	5	3	5	4	5	3	5	4	3	4	5	4	3	3	4	88
20	3	2	3	2	3	5	4	4	4	4	3	3	4	2	4	4	3	4	4	4	5	4	4	2	4	88
21	4	4	3	4	3	5	2	2	2	2	3	1	3	4	1	2	2	3	2	3	1	3	3	3	2	67
22	2	4	2	5	2	5	4	2	5	3	3	2	3	3	3	4	2	2	5	5	2	3	2	2	2	75
23	4	3	4	5	3	5	5	4	5	5	5	3	4	5	5	5	3	4	5	3	5	3	5	3	5	106
24	3	2	2	5	3	4	3	5	3	3	3	3	4	4	5	5	4	4	4	5	3	3	3	3	4	92
25	4	4	4	2	3	3	3	5	2	4	3	2	2	3	3	4	4	4	4	5	2	5	4	3	4	84
Total	68	80	68	76	68	94	73	76	80	82	80	57	69	84	87	86	81	74	84	88	73	82	73	64	75	1929

Rank the typewriters from least liked to best liked, and rank the raters from tough raters to easy raters.

One way to rank the brands of typewriters and raters is to use the sums of their scores, shown in the right-hand column and bottom row of Table 1. We can rearrange the entire table so that rows are arranged according to totals and similarly for columns. Table 2 shows the rearrangement of the rows and columns in this manner and also the row and column averages.

EXAMPLE 1.   ESTIMATING CELL VALUES FROM MARGINS

Use the marginal averages to estimate the cell values, taking account of the differences in brands and in raters.

<u>Solution</u>.   Although there are many ways to do this, some easy ways suggest themselves. One could just estimate every cell entry in a row by the average for the row. That wouldn't take account of the effects of raters. Or one could use the column averages as estimates of the cell entries in a column. But this would not take account of differences between brands.

One's next impulse is to estimate the cell value as the average of the row mean and of the column mean, but let us not be hasty. Let's think how these numbers might be built.

First, there is a grand average rating, let's call it $\bar{x}$, which equals 3.1. We use bars over letters to indicate that an average is being taken. A particular brand-rater pair departs from the grand average as a result of two effects: (1) the departure of the average score for that brand from this number $\bar{x}$, and (2) the departure of the average score for that rater from $\bar{x}$. For row i call $\bar{b}_i$ the brand average; for column j, call $\bar{r}_j$ the rater average. Then for the bottom row (brand 15) and last column (rater 6) $\bar{b}_{15} = 4.3$, $\bar{r}_6 = 3.8$. The gain from being brand 15 is

$$\bar{b}_{15} - \bar{x} = 4.3 - 3.1 = 1.2.$$

The gain from being rated by rater 6 is

$$\bar{r}_6 - \bar{x} = 3.8 - 3.1 = 0.7.$$

If these gains are additive, then the total gain is 1.2 + 0.7 = 1.9, which added to $\bar{x}$ gives 3.1 + 1.9 = 5.0. This model seems more reasonable than just averaging row and column means because ratings increase

Table 2. Typewriter ratings of Table 1 rearranged in order of row and column totals to show the basic orderliness of the ratings

Brand	12	24	5	3	1	13	23	21	7	18	25	8	4	10	2	17	22	11	19	14	16	15	9	20	6	Total	Avg.
7	1	1	1	1	1	2	1	1	1	1	1	1	1	1	1	1	1	1	1	1	1	1	1	1	1	26	1.0
13	1	1	1	1	2	1	2	1	1	1	1	1	1	2	2	3	2	1	1	1	2	2	2	1	1	38	1.5
4	1	2	1	1	2	2	2	1	1	1	1	1	1	2	2	2	1	1	2	2	3	2	3	2	1	39	1.6
18	2	2	2	2	1	2	2	1	1	1	2	3	2	3	2	1	1	3	3	3	3	2	1	2	4	49	2.0
8	2	1	3	1	2	2	2	3	1	3	3	1	2	3	2	2	2	3	4	4	4	3	4	2	3	62	2.5
6	3	2	2	2	2	2	2	2	2	2	2	3	2	2	4	2	4	3	3	3	3	3	3	2	3	63	2.5
14	3	3	3	2	2	2	2	2	2	1	2	3	4	3	2	3	3	3	2	4	4	2	4	4	3	66	2.6
21	1	3	3	3	4	3	3	1	2	3	2	2	4	2	4	2	3	3	2	4	2	1	2	3	5	67	2.7
16	2	3	3	4	3	2	4	4	2	1	4	1	3	2	4	2	2	2	2	3	3	4	4	4	3	72	2.9
22	2	2	2	2	2	3	2	2	4	2	2	2	4	3	4	2	3	3	5	3	4	3	5	4	5	75	3.0
12	3	2	3	3	2	2	3	3	4	2	3	4	2	3	4	4	3	3	3	2	2	3	4	3	5	75	3.0
3	2	4	3	2	4	4	3	3	2	4	3	3	3	3	3	5	3	3	3	2	4	3	2	4	5	81	3.2
25	2	3	3	4	4	2	4	4	3	4	4	5	2	5	2	4	5	2	4	3	4	3	4	5	3	84	3.4
11	3	3	1	3	1	5	3	4	3	4	2	3	4	5	4	4	4	5	2	1	4	5	5	4	4	86	3.4
9	2	1	3	4	2	1	4	3	4	5	4	4	3	4	4	3	3	4	5	5	3	5	4	4	2	86	3.4
1	2	3	3	3	3	4	2	3	3	4	4	2	3	4	3	5	5	3	5	3	3	5	4	4	4	87	3.5
20	2	2	3	3	3	3	3	5	4	4	4	4	2	4	2	3	4	4	4	2	4	4	4	4	5	88	3.5
19	2	3	3	3	3	3	3	3	3	4	4	4	5	3	3	5	4	4	3	3	3	4	4	4	4	88	3.5
24	3	3	3	2	3	4	5	3	3	4	4	4	5	3	3	4	3	5	2	5	5	5	3	5	4	92	3.7
5	3	4	4	3	4	2	3	4	3	3	3	4	4	3	4	5	5	4	5	5	4	5	4	5	5	94	3.8
10	2	3	3	3	3	4	4	4	5	2	3	4	3	4	5	4	5	5	5	4	3	5	3	5	4	96	3.8
2	3	3	4	4	4	3	4	4	4	4	4	4	3	4	5	3	5	5	5	5	5	5	3	4	5	100	4.0
17	3	3	3	4	4	5	4	3	4	5	3	4	4	4	3	5	3	3	5	5	5	5	5	5	5	101	4.0
23	3	3	3	4	4	4	5	5	5	4	5	4	5	5	4	3	3	5	5	5	5	5	4	3	5	106	4.2
15	4	4	4	4	3	3	4	5	5	4	5	4	4	4	4	4	5	5	5	5	4	5	5	4	5	108	4.3
Total	57	64	68	68	68	69	73	73	74	75	76	76	80	80	80	81	82	82	84	84	86	87	87	88	94	1929	
Avg.	2.3	2.6	2.7	2.7	2.7	2.8	2.9	2.9	3.0	3.0	3.0	3.0	3.2	3.2	3.2	3.2	3.3	3.3	3.3	3.4	3.4	3.5	3.5	3.5	3.8		3.1
Dev.	-.8	-.5	-.4	-.4	-.4	-.3	-.2	-.2	-.1	-.1	-.1	-.1	.1	.1	.1	.1	.2	.2	.3	.3	.3	.4	.4	.4	.7		

Rater

from left to right and also from top to bottom in Table 2. The proposed estimate is

$$\text{cell estimate} = \bar{x} + (\bar{b}_i - \bar{x}) + (\bar{r}_j - \bar{x}), \qquad (1)$$

which is one way of describing the basic structure of the table. The observed value in the cell is 5, and so the <u>residual</u>, observed value minus estimated value, is

$$\text{residual} = 5.0 - 5.0 = 0,$$

a close fit, surprisingly close, indeed misleadingly close.

<u>Exercises for Example 1</u>

1. Use Table 2 and the estimate (1) to compute the residual for the rating of brand 23, rater 20.

2. Use Table 2 and the estimate (1) to compute the residual for the rating of brand 7, rater 12. (Watch your signs.)

3. Using the estimate represented by equation (1), show that an equivalent estimate is

$$\text{cell estimate} = \bar{b}_i + (\bar{r}_j - \bar{x}). \qquad (2)$$

4. Use formulation (2) and the data of Table 2 to work exercises 1 and 2.

5. Interaction. When the residual is larger in absolute value than we think appropriate for the kind of errors that should be produced, we say there is an <u>interaction</u> between the row and column, here between brand and rater. That is, the effect produced is much more than can be accounted for by brand and rater effect plus error. Check the residuals of brand 11 and rater 14 and of brand 3 and rater 17 to see if you think these typewriters and typists interact unusually.

6. Write the algebraic expression for the estimate of the cell value which is the average of the row mean and of the column mean in terms of $\bar{x}$, $\bar{b}_i$, $\bar{r}_j$ and simplify. Compare this estimate with that given by equation (1). [Partial solution: New estimate is $\bar{x} + \frac{1}{2}[(\bar{b}_i - \bar{x}) + (\bar{r}_j - \bar{x})]$.]

7. Use the estimate of Exercise 6 and that of equation (1) in the text to compute residuals for a randomly chosen 10 cells in the table. Compare the two sets of residuals to try to see which estimate is doing the better estimating.

8.   Project. The Politics Table gives percentages of
        Frenchmen of various political parties saying they
        discuss politics with acquaintances. Rearrange
        the table so that row totals decrease from top to
        bottom and so that column totals decrease from
        left to right. Then compute the cell values based
        on row and column effects as discussed in the text.
        Compute residuals (observed value minus value esti-
        mated from row and column effects), and discuss
        what you find, both in the original table and in
        the residuals.

Politics table. Percent saying they discuss
politics in a French survey

Party preference	With family	With friends	With colleagues
Communist	68%	65%	69%
Moderates	53	53	36
MRP	51	43	40
Radical	58	49	31
RPF	58	51	32
Socialist	53	49	42

Source : Jean Stoetzel as quoted in Sidney
Verba's, Small Groups and Political Behavior:
A Study of Leadership, copyright (c), 1961,
by Princeton University Press, Princeton,
New Jersey (Princeton Paperback, 1972),
Table 1, p. 44. Reprinted by permission.

EXAMPLE 2.   SYSTEMATIC EFFECTS

Give evidence that the arrangements of ratings in Table
2 actually show systematic brand effects rather than
random fluctuations.

Discussion. This is not easy to do. If we had drawn
the arrangement of the numbers given in Table 1 from a
hat and then rearranged rows and columns as in Table 2,
the northwest corner would have low numbers and the
southeast high ones, similar to Table 2.
     Let us think of a feature of the table that sug-
gests that typewriters differ and then if we can, compute the probability of getting results at least as
extreme as the ones observed. Does some typewriter get
a large number of 1's so that raters agree that it is
poor? Yes, brand 7 got 24 1's. Of the 625 numbers in
the table, 74 are 1's, a proportion p = 0.118. What is
the probability of choosing 24 1's from such a population in a sample of 25?

If we draw with replacement it is (see reference [1])

$$\binom{25}{24}p^{24}(1-p)^{1} = 25p^{24}(1-p).$$

This number is practically zero, indeed about 1.17×10^{-21}, a decimal followed by 20 zeros. We would also have been impressed with either 24 or 25 1's and so we would sum over these two probabilities to find out how surprised we are that <u>this particular</u> machine got at least this many 1's. When we do, we get essentially the same tiny value, $P = 1.17 \times 10^{-21}$.

Of course, we didn't know that this particular machine would get so many low ratings. We should ask what the chance is that at least one machine got at least this many 1's. The chance that <u>no</u> machine got at least this many 1's, assuming independent drawings of ratings from the pot, with replacement, is

$$(1-P)^{25},$$

and so the chance that at least one machine got this many 1's is

$$1 - (1-P)^{25}.$$

When P is very tiny, this is approximately

$$1 - (1 - 25P + \text{small terms}) = 25P + \text{small terms},$$

and so the final result is approximately $25(1.17 \times 10^{-21}) = 2.92 \times 10^{-20}$, still very small.

This calculation is still not entirely fair because every table has some outstanding features and we may have chosen this feature of many 1's together because it was so outstanding in <u>this</u> table. For instance, why didn't we ask about the 5's instead of the 1's? Thus the probability we have chosen to compute is subject to further enlargement because of this possible selectivity. Unfortunately we don't know how much to enlarge it. If I suggest multiplying by 5 because there are 5 ratings someone will suggest that I might also have chosen to count numbers of 1's and 2's instead of just 1's. And so I might.

About all we can do is ask whether we can multiply our number by a large number and still have a small probability. In this instance we can. But as a general rule this is somewhat unsatisfactory, as after-the-fact calculations of probabilities are bound to be. It is a little like betting on a horse race after the race is over. Nevertheless, the kind of exploration we tried is a reasonable form of data analysis even if we can't properly assess the probabilities. Many 1's together is a reasonable form of agreement to inquire

about in our problem. See reference [2] for another
discussion of this problem.

One way around this difficulty is to have a few
standard, generally accepted methods for appraising the
systematic features of such tables. Then when we use
these methods, we are not so likely to be accused of
looking all over Robin Hood's barn for special features.
Some such methods, called the "analysis of variance"
have been developed and are widely used, but we shall
not take them up for this example. The weakness of
such standard methods may be that they do not bring out
the special features of the particular set of data we
have. And so exploration of data continues to be an
important art as well as science.

Comment. This analysis could have studied the probabilities of drawing without replacement

$$P(24 \text{ 1's}) = \frac{\binom{74}{24}\binom{551}{1}}{\binom{625}{25}}$$

$$P(25 \text{ 1's}) = \frac{\binom{74}{25}\binom{551}{0}}{\binom{625}{25}} \, .$$

However, when we ask the chance that at least one brand
randomly gets 24 or more 1's out of 25, the calculations become much more difficult. The answers we have
from our "with replacement" approach are quite adequate
here.

Exercise for Example 2

1.   In the manner of the text, inquire whether some
     brand has too many ratings of 5's to be explained
     by random assignments.

EXAMPLE 3.   RANDOM ARRANGEMENTS

Suppose that you wanted a random arrangement of the
numbers in Table 1 so that you could then form its
counterpart of Table 2 and compare the two. How could
you make one?

Solution. First you have to decide whether you want
the numbers drawn with or without replacement.
     With Replacement. Drawing a set of objects.
Suppose that the proportions of 1's, 2's, and so on
are as in the following table. The first column of the
table gives the rating, the second the proportion of
cells having each rating, the third the cumulative proportion, then the fourth tells what 2-digit random

numbers might be assigned to the rating. So if the
first random number drawn is 26, the rating assigned to
the upper left-hand corner cell is 2; if the next random number is 89, we assign a 5 to the second cell, and
so on. Naturally, this goes more rapidly on a high-
speed computer, once the program is written, than by
hand. But it can be done by hand if necessary.

Rating	Proportion	Cumulative proportion	Random numbers assigned
1	0.12	0.12	01 to 12
2	0.21	0.33	13 to 33
3	0.27	0.60	34 to 60
4	0.27	0.87	61 to 87
5	0.13	1.00	88 to 99 and 00

Without Replacement. Permuting or shuffling a set
of objects. What if one just wants to arrange these
particular numbers without replacement? One way would
be to get 625 cards of the same size, write 1's on 74
of them, since there are 74 1's in our table, and so on.
Then shuffle the cards thoroughly and deal them out.
This is hard to do. Thorough shuffling of 52 cards is
a serious effort, and shuffling 625 will be a great
deal of trouble.

To do the corresponding shuffling with random numbers, the following device will work. (Most devices
that come to mind do not work properly.)

Write the 74 1's in a column, and the correct number of 2's and so on, so that there is room for a 5-
digit random number beside it. Then assign to each
number in the column a 5-digit random number. Now rearrange the numbers in the order of the random numbers.
Let's rearrange a very small number of numbers this way
for illustration.

Original number	Random number	Rank of random number	Rearrangement of the original numbers
1	71562	6	2
1	21654	3	3
1	33021	4	1
2	67854	5	1
2	14406	1	2
3	20771	2	1
3	88942	8	3
3	73544	7	3

The arrows illustrate which number in the first column goes to what position in the last column. The last column is obtained by taking rank by rank the original number associated with it. Rank 1 was given to a 2, so it is first; Rank 2 was given to a 3, so it is second in the rearrangement, and so on. The last column is the shuffling of the first column determined by the random numbers.

We use 5-digit numbers to avoid ties. If a tie occurs, add further random digits to the right end for tied numbers until the ties are broken. We regard the 5-digit numbers as having decimals before them. Do not throw away ties and start over.

This would be a huge job for a table like Table 1, but we have carried it out on a computer so that you could see what the results would look like when the rows and columns were rearranged as we did for Table 2 for the real data. The results are shown in Table 3. You may wish to compare Table 2 and Table 3 in various ways to see how they differ.

Exercise for Example 3

1. Compare Tables 2 and 3 to see in what ways they differ. Discuss. Suggestions: How do the range of brand averages compare? How do the range of rater averages compare? Should you expect similar maximum and minimum averages for both brands and raters in Table 3 but not in Table 2? Why or why not? How do the grand means compare? How does the average of the 25 cells in the lower right corner compare with those in Table 2? How about upper left?

References

[1]   SBE, Weighing Chances, Set 4.

[2]   SBE, Weighing Chances, Set 1.

APPENDIX: TIES IN MAKING THE RANKINGS

Example 3 raised the possibility that two or more of the original numbers might draw identical random numbers. Is this likely? How large a number of digits should we use to keep the chance of ties occurring small?

If we use 2-digit random numbers to rank 5 objects by the method of Example 3, there is about one chance in 10 of getting at least one tie. We compute this by noting that the number of ways of getting no tie is $100 \times 99 \times 98 \times 97 \times 96$, that the number of possible

Table 3. Typewriter ratings of Table 1 randomly assigned and then rearranged in order of row and column totals

Brand	3	17	12	6	16	14	11	9	21	23	7	22	10	2	4	25	15	1	18	24	13	8	19	20	5	Total	Avg.
3	2	1	1	2	3	4	2	4	1	4	4	2	2	1	2	4	3	4	2	3	3	1	4	5	2	66	2.6
21	2	2	4	2	3	3	1	3	4	3	3	1	4	2	2	1	3	1	4	4	2	1	3	4	5	67	2.7
10	4	1	5	1	3	3	1	1	1	1	3	1	3	4	4	3	3	4	2	4	5	2	3	3	1	68	2.7
4	2	1	2	2	5	2	5	3	3	1	4	1	2	2	3	4	3	2	3	2	1	4	4	4	4	69	2.8
24	1	4	1	5	2	4	3	4	3	3	1	4	1	3	3	3	3	1	3	5	4	3	4	1	3	72	2.9
9	3	3	1	4	3	1	3	2	4	1	3	5	4	4	5	3	3	2	2	4	2	4	4	1	3	74	3.0
8	4	2	3	2	4	2	5	1	3	3	2	5	3	4	3	2	5	1	1	2	5	3	3	5	3	74	3.0
6	1	5	2	2	1	3	1	2	3	3	3	3	3	3	5	3	4	4	5	3	3	4	4	2	4	74	3.0
2	2	2	2	2	4	1	4	4	4	3	4	2	4	3	4	3	2	3	2	3	2	4	1	5	3	74	3.0
20	1	2	3	3	4	2	2	3	2	4	5	4	4	3	4	1	2	3	1	5	4	3	1	3	5	75	3.0
7	1	2	5	1	4	1	3	1	2	3	5	3	4	4	1	3	4	5	2	4	4	4	3	3	4	76	3.0
25	2	2	2	4	1	4	3	4	3	2	4	1	3	1	2	4	2	5	5	2	5	5	4	2	3	77	3.1
15	2	3	3	3	1	4	3	2	2	3	1	5	2	4	4	4	5	4	2	3	5	3	3	4	2	77	3.1
12	2	1	4	5	1	5	4	2	3	4	4	3	5	2	5	2	2	2	3	3	2	3	4	4	4	79	3.2
1	3	2	1	5	4	2	1	4	4	4	3	4	3	3	3	5	3	3	4	4	3	3	1	5	2	79	3.2
22	3	3	5	2	3	4	3	4	3	1	2	3	2	4	4	4	2	4	5	3	4	4	1	3	4	80	3.2
19	3	4	1	3	1	3	4	3	5	2	3	3	1	4	2	4	3	3	5	2	3	4	5	5	5	80	3.2
17	2	3	1	2	2	3	2	4	2	4	4	4	5	2	3	3	3	5	4	4	2	4	5	5	2	81	3.2
14	1	3	3	4	4	4	5	5	3	2	3	2	3	3	5	4	4	3	3	4	5	4	4	5	2	81	3.2
11	5	5	3	3	3	4	2	1	3	4	3	3	2	3	1	4	3	4	2	5	5	2	4	2	5	81	3.2
18	4	3	3	5	3	2	2	4	5	4	2	4	2	4	2	2	4	4	5	4	4	3	2	2	4	83	3.3
23	3	2	4	1	4	3	3	5	2	4	3	5	5	5	5	2	2	2	4	3	3	3	5	4	4	84	3.4
16	4	4	5	1	3	1	4	3	3	5	3	5	3	4	2	5	4	3	5	3	1	3	5	2	3	84	3.4
13	4	5	2	4	1	3	5	5	2	2	2	4	4	4	2	3	4	3	5	3	4	4	2	5	4	85	3.4
5	5	3	4	2	3	4	1	1	4	4	3	3	3	2	3	4	4	5	2	5	4	4	5	5	5	88	3.5
Total	65	70	70	70	71	72	72	72	74	76	76	77	77	78	79	80	80	81	83	83	83	83	84	88	88	1929	
Avg.	2.6	2.8	2.8	2.8	2.8	2.9	2.9	2.9	2.9	3.0	3.0	3.0	3.1	3.1	3.2	3.2	3.2	3.2	3.2	3.3	3.3	3.3	3.4	3.5	3.5		3.1
Dev.	-.5	-.3	-.3	-.3	-.3	-.2	-.2	-.2	-.2	-.1	-.1	-.1	.0	.0	.1	.1	.1	.1	.1	.2	.2	.2	.3	.4	.4		

ways to assign numbers is 10^5. The quotient is about 0.9, and so the probability of one or more ties is about 0.1.

On the average, how many ties will we get with 2-digit random numbers? Consider any pair, and assign the first one some random number. The chance that the second one gets the same number is 1/100. To get the average number of pairs tied when we rank 5 objects, we multiply by the number of possible pairs which is $5 \times 4/2 = 10$. This gives as the average number of tied objects $10/100 = 1/10$.

Is it a coincidence that the average number of tied objects approximates the probability of at least one tie? No, the average number of times a rare event happens is approximately the probability that it happens. To see this, let p_0, p_1, p_2, etc. be the probabilities that a rare event happens 0, 1, 2, ... numbers of times. Then $1-p_0 = p_1 + p_2 + p_3 + \cdots \approx p_1$ because if p_1 is small, then p_2, p_3 and so on are ever so much smaller--typically being like powers of p_1. The average number of successes is given by

$$0 \times p_0 + 1p_1 + 2p_2 + 3p_3 + \cdots .$$

The terms after the first, $2p_2 + 3p_3 + \cdots$, may add up to very little, if p_1 is small. Neglecting these later terms gives p_1 again.

Let us use the approximation of the average number of rare events for the probability in our problem. If we are ranking r objects and use N distinct random numbers, then the probability that any given pair have the same number is 1/N. There are $r(r-1)/2$ pairs and so

$$\frac{r(r-1)}{2N}$$

is the approximate probability of at least one tie, providing it is small, say less than 0.2. Obviously if $r(r-1)/2 > N$, the approximation is terrible because probabilities cannot exceed 1.

How many digits should we use in our random numbers? Or alternatively how many digits need we use to get a small probability of a tie. Let us say we want no more than one chance in 10 of a tie, and that we use k-digit random numbers. How big can r be if we want

$$\frac{r(r-1)}{2N} \leq \frac{1}{10} ?$$

Since $N = 10^k$,

$$r(r-1) \leq 2 \times 10^{k-1}.$$

We could now solve this quadratic in r, or notice that for good-sized r, $r(r-1)$ is approximately $(r - \frac{1}{2})^2$. Using this approximation

$$(r - \frac{1}{2})^2 \leq 2 \times 10^{k-1}$$

$$r - \frac{1}{2} \leq \sqrt{2} \times 10^{(k-1)/2}$$

$$r \leq \frac{1}{2} + \sqrt{2} \times 10^{(k-1)/2}.$$

When $k = 2$, $N = 100$, and $r < 5$. When $k = 3$, $N = 1{,}000$, and $r \leq 14$. Three-digit random numbers can rank only 14 objects with probability 0.9 of having no ties. Five-digit numbers similarly handle less than 142 objects.

What these results mean is that:

1.   We use large k to keep the chances of ties low, and plan to break the few that occur; or

2.   We program the computer not to assign a random number if it has already been used; or

3.   We use another method.

The problem treated here goes under the happy name of "the birthday problem" because it can be illustrated for everyday consumption by asking for the probability that at least two of a collection of people have the same birthday.

What we have just seen is that the delightful and astonishing result of that problem (23 people gives a probability of tied birthdays larger than $\frac{1}{2}$), leads to grievous difficulty in the genuine and important problem of making a random arrangement, even when we use a high-speed computer.

Exercises for Appendix

1.   Example 3 used $k = 5$, $N = 625$. Find the probability of a tie.

2.   How large should k be for the probability of a tie in Example 3 to be less than 1/10?

3.   If we wish the probability of a tie to be less than $1/10^t$ (t is not necessarily an integer), and use $N = 10^k$ random numbers, how small must r be?

References

[1] F. Mosteller, "Understanding the birthday problem," The Mathematics Teacher, 55 (1962): 322-325. See especially the left sides of equations 10, 11, and 12.

[2] F. Mosteller, R. E. K. Rourke, and G. B. Thomas, Jr., Probability with Statistical Applications, 2nd edition, Addison-Wesley, Reading, Massachusetts, 1970, pp. 97-98.

Collegiate Football Scores

FREDERICK MOSTELLER

INTRODUCTION

A well-organized table gives a surprising amount of information and suggests its own further analyses. Here we study a table whose entries record the results of all the collegiate football games played in the United States in 1967. From it arise such questions as: "If you know a team scored 14 points and yet lost the game, what was a likely score of the winning team?" "Do differences of scores tell us much about the two teams?" "Why should ties in football be about half as common as differences of one point?" Can the same kind of analysis be applied to baseball or hockey scores? The student learns to ask systematic questions of a complicated table. These examples are independent in the sense that each may be tackled without having to work through the others in the set.

SCORING IN FOOTBALL

The scoring in 1967 collegiate football allowed 6 points for a touchdown with a bonus opportunity to try for a 1- or 2-point conversion (1 point for kicking a placement after touchdown, or 2 points for a running play conversion after touchdown, the kick being much more frequently chosen), 3 points for a field goal, and 2 points for a safety, a very rare play. Since touchdowns followed by successful one-point conversions are frequent, we can expect many scores to come in

The author is at Harvard University, Cambridge, Massachusetts.

Table 1. Frequency distribution of 1967 college football scores

Winner (columns) vs. Loser (rows)

Loser	0	2	3	4	6	7	8	9	10	11	12	13	14	15	16	17	18	19	20	21	22	23	24	25	26	27	28	29	30	31	32	33	34	35	36	37
0	3		6		4	14	1	3	4		3	5	12	3	7	12		5	10	11	2	4	4	1	5	4	9	5	3	6		3	7	11	1	4
2					1	1									1					1						1										
3			1		1	5		1	1			3	4	1	1	1	1		3	2		1		1	1	1		1					1	1	1	
4																																		1		
6					1	11	1	1	1			5	6	1	3	5		2	1	5	1		4	6	1	4	8		4	3			5	3	2	2
7						3	1	7	12	1	5	8	16	1	2	9	3	4	10	14	3	7	8	1	2	6	10	3	8	3		2	2	3		2
8									2		2	4			1	1		2		2	1	1	3	1	1	1	1		1		1		1			
9								1	3		1		3						1	1		2	1						2	1			1			
10									1		2	4	10	3	2	2		1	4	6			1		2		2				3	1				1
11												1									1													1		
12												5	7			1	1	1	1	4	1	2	3				2		1	1			1	2		
13												4	12	3	2	7	3	1	2	4	2		3		3	4	3	1	3	2			1			1
14													4	3	4	8	1	4	10	12	1	7	7	1	4	7	11	1	2		3	2	6	6	1	1
15															1	3	1		4		1	1	2		1				1	1	1					
16															3	3	2		4	4	2		1	1		2	1		1					2		
17																			1	5				4	1	2	2						1	2	1	
18																		2	2	2					1								1	1		
19																			2	3							1	1	1	1			1	1		
20																				5	1	1	4		1	4	1	3		2	2		1			
21																				1	4	3	2		2	3	5	1	1			2	1	4		
22																						1	2	1	1				1	1	1		1			
23																						1	2											2		
24																								1	2	1	3			1	1	1				
25																														2						2
26																										1								1		
27																											2		1							2
28																												1	2	3		1	1			
29																													1			1	1			1
30																														1				1		
31																														1		1				
32																																1				
33																																				1
34																																		1		
35																																		1		1
36																																				
37																																					1
38																																				
Totals	3	0	7	0	7	34	3	13	24	1	13	39	74	15	27	52	12	22	55	82	20	35	53	8	27	42	61	18	30	35	9	15	30	45	9	20

Compiled from data in The World Almanac and Book of Facts, 1968 Centennial Edition, edited by Luman H. Long, published by Newspaper Enterprise Association, Inc., for the Boston Herald Traveler.

38	39	40	41	42	43	44	45	46	47	48	49	50	51	52	53	54	55	56	58	60	61	62	63	65	67	68	69	70	75	77	81	90	Totals
5		1	5	4	2	2			4	2	1	4	1				2	3	2	1		1	1			1	1			2	1	1	219
			1																														6
1	1							1																									36
																																	1
		2	4	4	2		1		1	1	1	3		3	1			3		1											1		114
2	1		6	1	1		1	2	3	2	1			1		2	1	1	1		2			1			1						186
	2	1	1								1															1							32
				2																													19
	1			2																													48
					1																												4
			1	1						1						1		1										1	1				42
	2			2				1	2	2	1				1																		72
2	1		2	5	1	2	1			2	1	1		3							1												128
2			1		1													1															23
		1	1	1														1															29
			1	1			1	1																									23
2					1				2																								14
1				1		1																											14
1							1				1	1																					29
			1	2	2			1								1																	36
			1	1		1	1		1																								14
			1										1					1															8
1	1							1																									13
1																																	5
																																	2
			2	1		1	1																										10
		1		1								1																					11
																																	4
																																	2
	1																																3
					1						1																						3
																																	1
		1																															2
																																	2
																																	0
				1																													2
		1																															1
18	10	7	28	30	12	7	7	6	13	12	7	6	5	8	2	4	3	10	4	1	4	1	1	1	1	1	2	1	1	3	1	1	1158

multiples of seven, or nearly that. Field goals, while
frequent, are not as frequent as in professional football, partly because the goal posts are 10 yards
farther away in the collegiate game.

Tables 1 and 6 give the joint frequency distributions of winning and losing scores for collegiate football games in 1967 and 1968.

EXAMPLE 1. PREDICTING WINNER'S SCORE FROM LOSER'S--
REGRESSION METHODS

From the data of Table 1 find how the winner's average
score changes as the loser's score increases.

Solution. To relate the winner's score to the loser's,
let us compute the mean winning score for each losing
score as far as there are enough games to make the calculations reasonably reliable. For example, for a
losing score of 25, the average winning score is

$$\frac{2\times31 + 2\times37 + 1\times38}{2 + 2 + 1} = \frac{174}{5} = 34.8$$

and so on.

Table 2 shows the results. Its first row gives 0
as the loser's score in a total of 219 games, and an
average winning score of about 27 points (26.7). The
corresponding calculations made for each of the rows
having at least 10 games, plus results obtained by
pooling the results from 23 to 33 inclusive and
plotting at 28 for losers and 34.7 for the winners,
give the dots plotted in Fig. 1. These numbers give
predictions of winning scores for each losing score.
For successive sets of 5 losing scores (0-4, 5-9, etc.)
we have computed the average winning score and the
average losing score. We then plotted a + point to
summarize these averages.

We have drawn a freehand smooth curve through
these + points to represent an idealistic smooth fore-
cast of winner's scores from the loser's score by
groups of 5 losing scores. Such a curve is called the
regression curve of the winning scores on the losing
scores. This curve provides a baseline against which
to assess departures of the single points. By looking
at the departures of the points from the curves--the
residuals--we can tell whether some scores contain
special information. In particular, a score of 0 by
the loser means that the team is scored on more frequently than the fitted curve would indicate--that is,
that the average team with 0 score is slightly more
scored upon than the mere zeroness of their score might
lead one to expect. This idea, but with the effect re-
versed, is even clearer when we move up to a score of 3
points for the loser.

Table 2. Average winning score for each losing score

Losing score	Number of games	Average winning score	Losing score	Number of games	Average winning score
0	219	26.7	20	29	28.9
2	6	19.7	21	36	30.8
3	36	19.0	22	14	33.6
4	1	35.0	23	8	36.1
6	114	27.8	24	13	31.3
7	186	24.5	25	5	34.8
8	32	25.1	26	2	31.0
9	19	21.8	27	10	37.4
10	48	20.4	28	11	34.7
11	4	28.5	29	4	33.8
12	42	27.0	30	2	33.0
13	72	24.3	31	3	34.3
14	128	27.8	32	3	41.7
15	23	27.4	33	1	36.0
16	29	24.6	34	2	38.5
17	23	29.2	35	2	36.0
18	14	30.6	37	2	39.5
19	14	29.6	38	1	40.0
			Total	1158	26.7

Note that for a loser's score of 3, the mean winning score falls well below the freehand curve. This suggests that a team that has scored one field goal can be thought, on the average, to be showing additional competence relative to the other team in defense--about $6\frac{1}{2}$ points worth. A score of 6, usually reflecting a touchdown without extra point, is a slight indication of less competence. No doubt the implication is three-fold, not being able to kick the extra point is itself a sign of less competence, or having a kick blocked is a sign of extra competence on the part of the other team, or needing a 2-point conversion and failing to make it leaves the other team obviously ahead. And so any way one looks at it, a score of 6 by the loser suggests better scores for the winner than the smooth curve would predict.

One might then expect that a score of 7 on the part of the loser would make them look a little better, and relative to a score of 6 they do perform better, but they fall just about on the smooth curve. A losing score of 9 or of 10 shows a good deal of strength for the loser, or shows that the teams are more evenly matched than the smooth curve implies, again perhaps because of the field goal, or because the field goal was worth trying for.

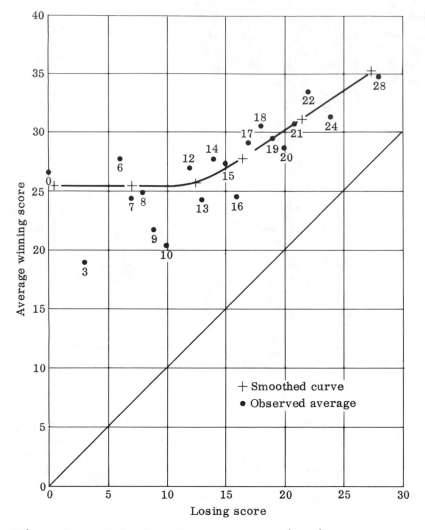

Fig. 1. Graph of average winning score for each losing score. The number attached to a dot is the losing score.

Exercises for Example 1

1. Note that the curve of Fig. 1 generally slopes upward. Why?

2. For each losing score arising in 20 or more games, in Table 1, find the median winning score. (The median score is the score in the middle when scores are lined up from least to greatest. For example, the median of 7, 9, 10, 10, 12, 14, 14 is 10; and the median score of 7, 9, 10, 10 is midway between 9 and 10.)

Use ordinary graph paper to plot these medians vertically against the losing score in the horizontal direction, using the same scale on both axes.

3. Give an interpretation to the relation between the points for the losing scores 0, 3, 6, 7, 10.

4. As the losing team scores more, is the difference (winner's score - loser's score) generally increasing or decreasing?

5. Project. Make a graph like Fig. 1 for the baseball scores of Table 3.

Table 3. Frequency distribution of baseball scores for the American League in 1968

Losing score	Winning score																Totals
	1	2	3	4	5	6	7	8	9	10	11	12	13	14	15	16	
0	38	35	28	20	8	7	8	2	3	2		2	1				154
1		66	42	42	28	17	7	9	3	3	1	2	2				222
2			61	41	21	14	20	8	3	7	3	2			1		181
3				56	24	20	13	7	3	2	3	1		1			130
4					29	11	6	10	1	2	1		1				61
5						13	6	4	3	1		1					28
6							13	2	2		3	1					21
7								2	2								4
8									1	2		1	1				5
9										2							2
10											1		1				2
Totals	38	101	131	159	110	82	73	44	21	21	12	10	6	1		1	810

Source: Scores from Official Baseball Guide for 1969. Chris Roewe and Paul Macfarlane, editors; C. C. Johnson Spink, publisher. Published by The Sporting News, St. Louis.

EXAMPLE 2.   PREDICTING LOSER'S SCORE FROM THE WINNER'S

Given the data in Table 1, find for a given winning score the average losing score.

Solution.   Before going further, try to guess the shape of the graph that relates the average losing score to each possible winning score. It is hard to guess. We construct Table 5 from Table 1. For example, given the winning score 3, the average losing score is $(6 \times 0 + 1 \times 3)/7 \approx 0.4$, and so on. Then plot the points (average losing score, winning score) to get the graph in Fig. 2. We have drawn a freehand curve through the points. Its peculiar shape is easily explained after it is seen, but few people think of it in advance. The curve must start with the "loser" having mean 0 because if the "winning" team scores 0, the "losing" team must also, according to our rules. Then as the winning team scores more and more, there is room for the losing team to get higher scores and it does for a ways. But this trend finally reverses itself, presumably because when one team runs up a high score, this is likely a signal that the teams are not well matched and that one of the teams has the ball nearly all the time. And so both these reasons suggest that the losing team may score less freely. The curve makes these remarks quantitative. It is best to look at this curve with the winning score axis horizontal and the losing score vertical. Then it corresponds to looking at Fig. 1.

We cannot read so much into specific values of the winning scores as we did into the losing scores, partly because they have fewer games available per score, and so the numbers are less well determined. Also the structure of the winning score is harder to deduce from its number--30 can be scored in many different ways, some suggesting less competence, others suggesting more.

The regression curve in this example is called the regression of the losing score on the winning score. The two regression curves on Examples 1 and 2 differ a great deal, because the predictions they are making are very different.

Exercises for Example 2

1.   Approximately what winning score has maximum average losing score?

2.   Use the baseball scores of Table 3 to make a graph like Fig. 2, the regression of losing scores on winning scores.

EXAMPLE 3.   DIFFERENCES IN SCORES

Given Table 1, get the graph of the frequency distribution of differences between scores and discuss the results.

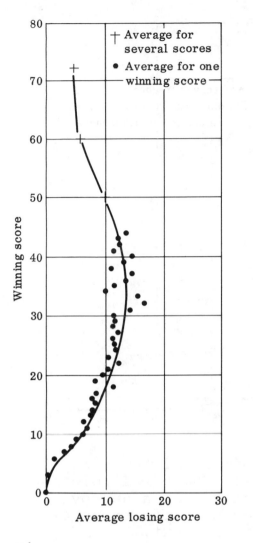

Fig. 2. Graph of average
losing score for each
winning score.

Solution. To get the frequencies of the differences we
must sum along southeastern diagonals. The results of
these sums are shown in Fig. 3. The horizontal axis of
Fig. 3 shows the difference d between the winner's and
the loser's score, the vertical axis the frequency of
that score in games. Within each block of 10 scores
(0-9, 10-19, etc.) we put a + at the median--the number
midway between the two middle numbers in the block,
both horizontally, and vertically. Then a freehand
smooth curve was drawn through these medians. Some
counts which stand especially far from the curve either
above or below have been boxed as before. We say that
these points have large residuals. These numbers may
have some special significance over and above the
relation implied by the curve.

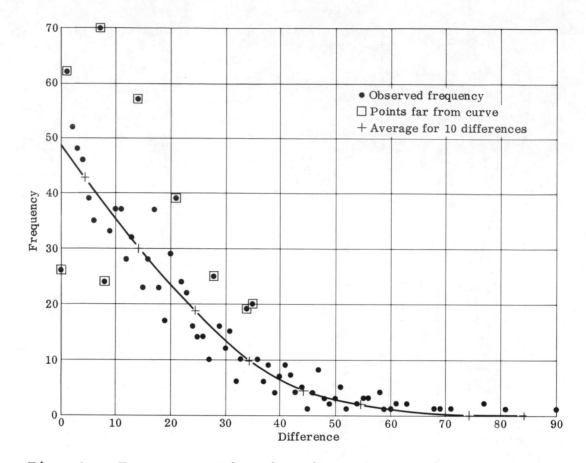

Fig. 3. Frequency distribution of the differences
between winning and losing scores. The boxed points
fall far from the curve. Most can be explained by
"touchdown scoring", but the cases of differences of
0 and 1 are studied especially in the text.

Discussion. A difference of 0 seems to occur too sel-
dom if the fitted curve is to be believed. Why would
this happen?

     To understand the reason, it helps to think of
close scores and named teams instead of winners and
losers. When Harvard and Yale play, a score 14-13 can
happen two ways, but a score of 14-14 can happen only
one way. We would guess that scores very close to one
another would be nearly equally frequent, except for
lumpiness. Thus if we think of teams A and B as nearly
equally matched and of the scores being assigned to
them in the order A, B, then we think of 14-13, 14-14,
13-14, and 21-20, 21-21, 20-21 and so on as each
forming 3 nearly equally likely game scores. When we
assemble all these sets, then we should find that the
ties happen about half as often as the 1-point differences. Some may worry that the 14-13, 14-14, 13-14

idea neglects the possibility of a 13-13 tie also.
But to check this out, one needs only think of a large-
size checkerboard, as in Diagram 1, to see that there
are two diagonals, one above and one below the main
diagonal, and each almost as long as the main one.

Team A's score

		0	1	2	3	4	5	...
Team B's score	0	tie	d=1					
	1	d=1	tie	d=1				
	2		d=1	tie	d=1			
	3			d=1	tie	d=1		
	4				d=1	tie	d=1	
	5					d=1	tie	
	⋮							⋱

Diagram 1.   Checkerboard layout suggesting that
ties are about half as frequent as differences
of one point.

     This argument makes it reasonable to expect that
ties should happen only about half as often as scores
differing by 1 point, or even by 2 points. Consequently, we should look to see whether the number of differences of 1 is about twice the number of differences of
0. There are 62 differences of 1 and 26 differences of
0, so when we double the ties, we are still left with
an excess of 10 differences of 1. It is this difference that needs explanation, not the whole difference
between 62 and 26. One possibility is that when the
teams are not evenly matched, differences of 1 become
more likely than differences of 0, but another might be
the popular attitude of the players and fans that a
team should try to win rather than to tie. A third is
that the difference between 62 and 52 may be largely
owing to year-to-year sampling fluctuations. Indeed,
theory not given here suggests that differences this
large or larger would occur nearly half the time on the
basis of random fluctuations. Note that a difference
of 2 did occur 52 times, just twice the 26 zeros.

Exercises for Example 3

1. Use the 1968 collegiate football scores (Table 6) to check whether the frequency of ties is about half the frequency of differences of 1.

2. Use the hockey scores of Table 4 to see whether ties are about half as frequent as differences of 1. Discuss.

3. Project. Make a graph like Fig. 3 for the baseball scores of Table 3. See if any differences stand out.

4. Project. Make a graph like Fig. 3 for the hockey scores of Table 4. See if any differences stand out.

Table 4. 1967-1968 Professional hockey scores

| Losing score | Frequency distribution of scores Winning score | | | | | | | | | |
	0	1	2	3	4	5	6	7	8	9	Total
0	2	14	16	16	10	5	7	2	1	0	73
1		14	26	42	22	17	9	4	2	1	137
2			28	28	38	16	9	6	1	2	128
3				22	14	22	9	1	0	3	71
4					6	5	7	5	0	1	24
5						3	1	3	1	0	8
6							1	0	2	0	3
Total	2	28	70	108	90	68	43	21	7	7	444

Source: Scores provided by Ron Andrews, Publicity Director, National Hockey League.

Table 5.   Average losing score for each winning score

Winning score	Number of games	Average losing score	Winning score	Number of games	Average losing score
0	3	0	28	61	11.2
3	7	.4	29	18	10.6
6	7	1.6	30	30	10.6
7	34	3.1	31	35.	14.4
8	3	4.3	32	9	16.8
9	13	5.2	33	15	15.9
10	24	6.1	34	30	10.1
11	1	7.0	35	45	11.6
12	13	6.2	36	9	13.8
13	39	7.4	37	20	14.8
14	74	7.9	38	18	11.1
15	15	8.5	39	10	13.1
16	27	7.8	40	7	14.7
17	52	8.3	41	28	10.8
18	12	11.3	42	30	12.8
19	22	8.3	43	12	12.6
20	55	9.6	44	7	13.7
21	82	10.4	45	7	16.1
22	20	12.4	46	6	12.5
23	35	10.5	47	13	8.2
24	53	12.0	48	12	9.7
25	8	11.8	49	7	13.1
26	27	11.3	50	6	13.3
27	42	12.1	⋮	⋮	⋮
			Total	1158	10.0

Discussion of differences continued. Aside from differences of 0 and 1, we note that the frequencies of
differences of 7, 14, 21, 28, 34, and 35 lie well above
the curve, presumably reflecting the 7-pointness of the
touchdown plus 1-point conversion. We are encouraged
to see the routinely obtained pattern of residuals
emphasize the special nature of multiples of 7 in a
problem where we know that such special effects should
occur. It makes us hope that other patterns of residuals standing out in other problems will indicate real
effects in the process generating the data. Points
falling below the curve are a bit harder to identify,
but 8, 19, 27, and 32 seem to fall below, although only
8 is especially convincing. One might have thought
that 14-6 would be a frequent score that would produce
many 8's, but that did not happen.

EXAMPLE 4.   AVERAGE SUM VERSUS DIFFERENCE

Given Table 1, plot the average sum of scores against the difference of the scores. Discuss the shape.

Solution. The scores with fixed differences fall along southeastern diagonals. We need to compute the average sum for each of these. Then we have averaged these sums for the difference between 0 and 9, 10 and 19, and so on. We have also obtained the average difference in the same interval and plotted (average difference, average sum) as the dots in Fig. 4.

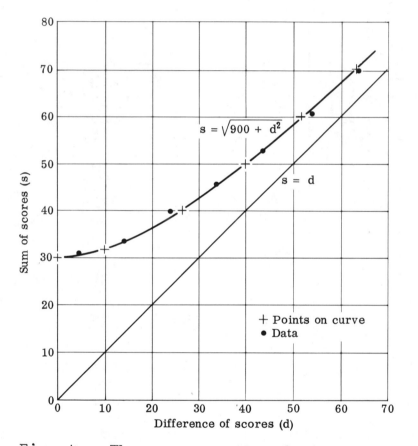

Fig. 4. The average sum of scores for a given difference. A hyperbola is fitted to the data.

These dots looked as if they might fall along half of a branch of a hyperbola. A hyperbola in s and d might have the form

$$\frac{s^2}{a^2} - \frac{d^2}{b^2} = 1.$$

However, if, as the data suggest, it is to come close to the 45° line s = d when d grows large, then a = b.

We need then a value for a in the equation

$$s^2 - d^2 = a^2.$$

When d = 0, we observe from Fig. 4 that s = 30. And so an equation to try is

$$s^2 - d^2 = 900 \tag{1}$$

or

$$s = \sqrt{900 + d^2}.$$

When d = 40, s = $\sqrt{900 + 1600}$ = 50, which fits the set of dots well. The + points are computed from equation (1). We have no theory to suggest a hyperbola, let alone this one. For the football scores, the fitted curve is just one summary of the data.

Exercises for Example 4

1. From Fig. 4, use the fitted equation to check the value of the ordinates when d = 40.

2. From consideration of football games, why might s be approximately equal to d for large d?

3. Project. Use the hockey scores of Table 4 to produce a graph like Fig. 4. Can you fit the data well with a hyperbola?

4. Project. Do Exercise 3, using the baseball scores of Table 3.

Reference

[1] For additional discussion of the football scores, see SBE, Exploring Data, Set 7.

Table 6. Frequency distribution of 1968 college football scores

Losing score \ Winning score	0	2	3	4	5	6	7	8	9	10	11	12	13	14	15	16	17	18	19	20	21	22	23	24	25	26	27	28	29	30	31	32	33	34	35	36	37
0	5		2			4	4		3	5		2	2	8		6	7	2	3	4	7		1	6	2	6	2	12	3	2	8	1	2	5	6	3	3
2							2																														
3						3	4		1	3		1	1	2			2	1	2	2	2			1		2		1	1	1	3		1		1		
4																																					
5																								1													
6							10			4		2	2	6		3	4	1		5	7	1	2	1		2	4	3	1	2	3	4	2	2	6		2
7							5	3	3	6		1	6	14	1	7	4	3		4	12	4	4	7	1	2	6	9	5	4	6	2	2	4	3		5
8										4			3			2			1				1			2		1		1	1		1	2		1	
9										4		1	3	1		1				1		2	1	2	1		1	2			2	2		1	1		
10										2	1	2	6	5			6			6	2			1	1	1	1	2			1		2	1		1	
11														1							1			1		1	1			1	1						
12													1				5			3	5				1		2	8		1	2		1	1	3	2	1
13													2	4	1	2	7	1	3	4	4	3	4	4	1	1	7	6		2	1	1		2	6		3
14														5	1	4	7	2	1	7	8	1	5	7	1	2	7	10		4	8	3	1	3	5	2	3
15																1	3	1		2	1		1		1		2	3				1		1	2		1
16																	2			2	3		2					1		2			1	1	1		1
17																	1	1	2	6	5	1		3		1		2	2				1	1	4		
18																			1				1					2		1	1						
19																					3	2	2	2		1	2	1			2			1	1		
20																				2	9		5	3	2	3	7	2	1	4	2		1	1	4		
21																					1	4	3	4	1	1	1	2		2	1			1	1		
22																								1	1	1	1	2		1	1			1			1
23																										2	1	2					1			1	
24																								1	1	1	3	1	2		1	1	2				
25																											1				1			2	1		
26																											2		2								
27																												2			1	2	1	2	1		
28																												2		3	2			2	2		2
29																													1						1		1
30																																					
31																																	1	1			
32																																		1	2		
33																																			1		
34																																			1		
35																																			1		
36																																					
37																																					
38																																					
40																																					
42																																					
45																																					
47																																					
48																																					
Totals	5	-	2	-	-	7	25	3	7	28	1	9	26	46	3	26	48	12	13	48	70	18	32	45	14	29	51	76	18	31	48	17	20	36	54	10	23

Compiled from data in The World Almanac and Book of Facts, 1969 Edition, edited by Luman H. Long, published by Newspaper Enterprise Association, Inc., for the Boston Herald Traveler.

38	39	40	41	42	43	44	45	46	47	48	49	50	51	52	53	54	55	56	57	58	59	60	61	62	63	64	65	66	68	69	71	76	77	100	Totals
3	2	4	4	2	1	1	1	4	3	3	3		1	2						1	1				1	1		1	1	1	1	1			169
1	1			1						1		1													1								1		2
																																			-
																																			1
1	1		2	3	1	3	1	1	1	2	1						1			3						1			1					1	102
3		2	1	6	4	1	3			1	2	1				1	1	1							1				2						163
1	1					1	1			2											1			1		1									28
																										1									27
														1																					42
																																			7
1			1	2					1																										41
		1	4	2	1		1		1	1						1	1	1	1																84
3			6	2	1	2	1	3	4	1	3	1																		1					125
1	1					1			1		1																								25
						1			1	1																									19
			1	2								1																							34
	1	1		1						1	1											1													12
			1	2					1			1									1	1													24
1		2		1				2	1		1	1					1							1					1						59
			1	2			5	1	3								1					2													37
	1	2	1	1																															15
1				1				1																											10
				1				1									1																		15
								1											1	1															8
			1	2					1																										8
	1	1							2																										14
				2			1				2	1				1									1										21
1	2																					1													7
								1						1																					2
	1								1																										4
																1																			4
																														1					2
2																2													1						6
1																					1														3
		1													1																				2
1																																			1
															1					1															2
					1							1																							2
												1																							1
																	1																		1
																													1						1
																							1												1
21	10	17	22	34	9	10	13	14	20	16	12	9	2	4	3	2	7	5	3	9	4	3	1	2	4	1	2	1	6	2	1	1	1	1	1173

Periodicities and Moving Averages

FREDERICK MOSTELLER

INTRODUCTION

Sometimes when we look at data that come in a sequence, it seems as though the numbers go up and down with some regularity. But often there is no such regularity until we study the data more closely, and even then, what appears to be a regularity may only be the result of the way we study the data. Our aim here is to make you aware of how the method of analysis can bring into the data an apparent regularity which was not originally there.

The search for periodicities is one of the most useful and yet most tricky tasks in the analysis of data that arises from observations taken regularly in time. Typical of these data are stock prices, gross national product, population figures, sunspots, rainfall, tide levels, egg production, and electronic impulses. If one found a periodicity in stock prices, one could buy at the low, sell at the high, and become a philanthropist.

EXAMPLE 1.

Many kinds of time series have natural variation in them that makes the main trend in the data hard to see. This variation is called "noise". To see an example of a time series that is made entirely of "noise", look at Fig. 2. These points are generated by a random process that has mean 22. All you see in the picture are

---

The author is at Harvard University, Cambridge, Massachusetts.

1

random fluctuations about that value. Your eye, like mine, may be deceived by what looks like a slight trend from low to high as the eye moves from left to right. It is pure chance.

In such series, to get a better grip on the measurement at a given point, one is inclined to average several measurements close to one another. For example, we might sum (or average) 5, then throw away the first and add the next and sum these. (The choice of 5 is arbitrary.) Here is an example.

Time period	1	2	3	4	5	6	7
Raw observations	6	5	8	0	6	2	2
Moving total					25	21	18

In this example we have chosen to use random numbers initially because we need to start with an artificial example where we can be confident of having no trend of any kind. We use the random digits 0,1,2,3,4, 5,6,7,8,9 from a random number table as our raw observations. We extended the numerical example just given to 45 such sums of overlapping random numbers. Each sum is plotted in Fig. 1. Look now at Fig. 1.

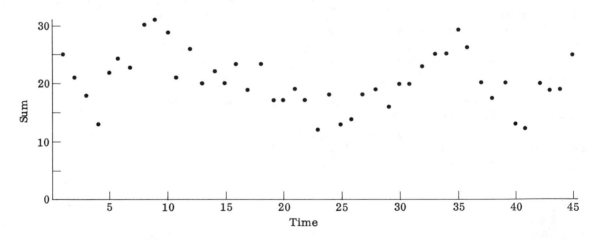

Fig. 1. Moving sum chart. Each point is sum for 5 periods.

1. Approximately what is the average of these sums (by eye)?

2. Sketch lightly with a pencil the main trend in the data. Then count the number of times this trend crosses to get a notion of the period of this gross curve.

3. Find out how often successive points of the original data in Fig. 1 cross the 20½ line. Get the average time to cross. (I choose 20½ to avoid the ties that would be caused by 20.) If one point is

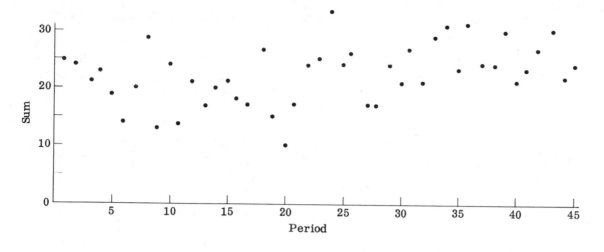

Fig. 2.   Sums for 5 successive periods, nonoverlapping.

above 20½ and the next below 20½, that is a crossing. If the next point is above 20½ that is another crossing.

4.   Compare this graph with a comparable one made by adding up 5 nonoverlapping random numbers 45 times from your random number table. Make a similar analysis and discuss the results.

Solution

1.   There are 19 points above and 19 below the 20 line, and so the average sum is about 20. The <u>median</u> is exactly 20. It is the number that has no more than half above or below it.

2.   Different people will have different sketches. Mine crossed the 20 line at times 2.5, 4.5, 15.5, 30, 37.5, 44.5. Consequently between 2.5 and 44.5 there were 5 half periods with average length 8.4 time units. So the cycle length is about 17 time units.

3.   12 crosses of the line or a successive point crossing rate of 45/12 ≈ 3.7.

4.   Figure 2 shows the results of summing <u>nonover-lapping</u> sets of 5 random digits.

a)   Since we are using the sum of 5 of the random digits 0,1,2,3,4,5,6,7,8,9, which average 4.5, the mean should be 22.5 (= 5 × 4.5).   21 points are below this value, which is close to half of the 45 points.

b)   The graph of Fig. 2 seems to have only one dip and rise, as far as big trends go.

c) The count of successive points crossing 22.5 is 20.

d) Overall the nonoverlapping graph has less long-term periodicity and slightly more short term crossings. Figure 2 seems the more random.

Discussion. The series shown in Fig. 1 was constructed from the same sort of series as that for Fig. 2--random numbers--but it used the moving sum technique. The evidence of these graphs suggest that the method of moving sums actually creates periodicities even when the original numbers are completely random. This astonishing suggestion turns out to agree with a mathematical fact; this periodicity, artificially created by the analysis, has been named the Slutsky-Yule effect after investigators who studied it.

Exercises for Example 1

1. Make another set of moving sums based on 5 successive periods, using random numbers, and see whether it behaves generally like the points in Fig. 1.

2. Make a graph like that for Fig. 1, based on moving sums of 10 random numbers, and see whether the periods seem longer.

You may already have figured out why this effect happens. Since the original numbers vary, the sums are bound to fluctuate. But when a moving sum is larger than the median, then it has a restoring force pressing it toward the median. Why? If the sum is high, the next number is likely to be smaller than the average number making up the sum, and therefore smaller than the one that is to be dropped. It may be instructive to think about all possible moving sums of 3 numbers when only the three digits 0, 1, 2 are used. Then the average sum is 3. The sequences in the first column add to more than 3. Their totals are shown in the second column. The third column tells how many of the digits are less than the item to be dropped, the fourth the fraction of the time each total leads to a smaller number after replacement, and the other four columns give the corresponding information for "more" and for "equal". (See table on the next page.)

Note that of the 30 possible outcomes for the next total, 1/2 would send it down, and 1/3 would leave it the same; only 1/6 would send it up. The higher the total, the more likely it goes down, rather than up.

Next digit to be dropped ↓	Total	Numbers less than item to be dropped		Numbers larger than item to be dropped		Numbers same as item to be dropped	
222	6	2	2/3	0	0/3	1	1/3
122	5	1		1		1	
212	5	2	5/9	0	1/9	1	3/9
221	5	2		0		1	
022	4	0		2		1	
202	4	2		0		1	
220	4	2	8/18	0	4/18	1	6/18
211	4	2		0		1	
121	4	1		1		1	
112	4	1		1		1	
		15		5		10	

3.   Make the corresponding table for sums less than 3 and discuss the chances that the next sum will be higher.

EXAMPLE 2

Now that we have seen artificial periodicities created by moving averages in data free of any trend, let us see an illustration in data from real life. Figure 3 shows daily lowest temperature at the Boston Airport for the first 51 days of 1969. Figure 4 shows 51 5-day moving sums and Fig. 5 shows 51 10-day moving sums. Note how much smoother and more wavelike the movements in Figs. 4 and 5 are compared to Fig. 3. Whereas the raw data in Fig. 3 shows 13 crossings of its median (23), the 5-day moving sums show only 8 crossings and the 10-day sums show 7.

The moral here is that taking sums or averages smooths the data and can give a more reliable answer at a particular point, but the resulting pattern of points can be misleading in presenting such a clear picture of regular smooth waves.

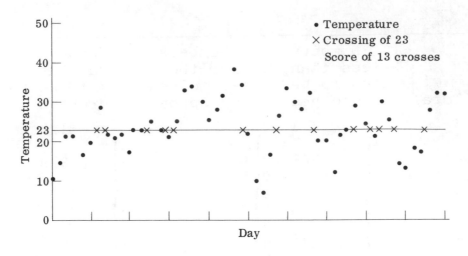

Fig. 3.   Lowest daily temperature, Boston Airport, January 1 to February 20, 1969.

Source: <u>Climatological Data</u>, 1969.   U. S. Department of Commerce, Maurice H. Stans, Secretary; Environmental Science Services Administration, Environmental Data Service.

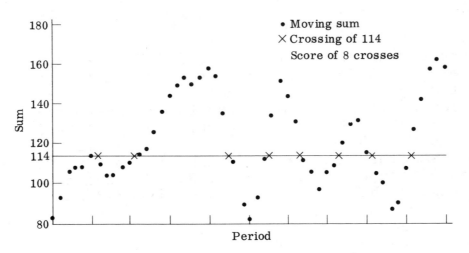

Fig. 4.   Moving sums of lowest daily temperature, Boston Airport, January 1 to February 24, 1969; 5-day periods.

Source: <u>Climatological Data</u>, 1969.   U. S. Department of Commerce, Maurice H. Stans, Secretary; Environmental Science Services Administration, Environmental Data Service.

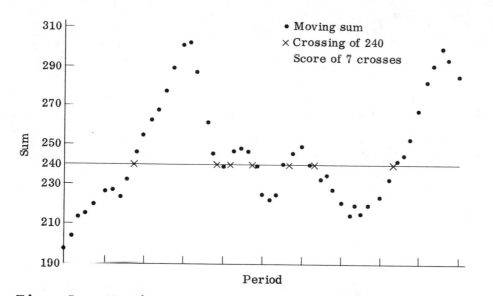

Fig. 5.   Moving sums of lowest daily temperature, Boston Airport, January 1 to
March 1, 1969; 10-day periods.

Source : Climatological Data, 1969. U. S.
Department of Commerce, Maurice H. Stans,
Secretary; Environmental Science Services
Administration, Environmental Data Service.

Prediction of
Election Results
from Early Returns

JOSEPH SEDRANSK

INTRODUCTION

Predicting election results from samples of early returns is an increasingly popular procedure for many TV
and radio stations. This set studies sampling techniques which make reasonably reliable predictions long
before most people would suppose that the election had
been settled. Prediction of results by such techniques
is compared with the final results in an actual election in Iowa in 1964.

The national television and radio networks spend
large sums of money to use early returns for accurate
predictions of the final outcomes of election contests.
During the evening, however, local stations often report only the latest voting percentages; for example,
the current Democratic vote divided by the current
total vote. Because this often does not accurately
reflect the final percentages, several faculty members
at Iowa State University decided to try to improve the
accuracy of election predictions from early returns
even when only a very limited amount of money and time
were available. Both the 1962 and 1964 elections in
Iowa have been investigated [1, 2] and the results are
encouraging.

In 1964, one goal was to predict in the Iowa election for governor the percentage of the final vote to
be obtained by the Democratic party's candidate, Harold
Hughes. At any time, t, during the evening, the

The author is at the University of Wisconsin, Madison,
Wisconsin.

1

information available to us for each of Iowa's 99 counties was

1. the total number of precincts reporting their vote at time t,

2. the total Democratic party vote for the precincts in (1), and

3. the total vote for the precincts in (1).

In addition we had some past voting records and current population information.

The current vote percentage, which we will call "the uncorrected vote percentage" and denote by $\hat{P}_u$ ("u" for "uncorrected") is, at any time t, the total Democratic party vote d in the state divided by the total vote w in the state; i.e., $\hat{p}_u = \frac{d}{w}$. (The circumflex "^" indicates an estimate rather than the true value.) The vote comes in county by county; thus,

$$\hat{P}_u = \frac{\sum\limits_{c=1}^{99} d_c}{\sum\limits_{c=1}^{99} w_c}$$

where d_c and w_c represent, respectively, the Democratic and total vote in the cth county and there are 99 counties in the state. (Note $\sum\limits_{c=1}^{99} d_c$ is mathematical shorthand for $d_1 + d_2 + d_3 + \ldots + d_{99}$ and similarly for $\sum\limits_{c=1}^{99} w_c$.)

While $\hat{P}_u$ is the estimate reported by many local stations, it may well be, and frequently has been, a poor estimate of the final percentage voting Democratic.

Before reading on you might try to think of several reasons why this might be so.

An important source of this difficulty is that, with early returns, certain segments of the electorate may tend to be represented disproportionately. For example, many precincts in cities use voting machines and they may be overly represented among the early election results. If such precincts give a much larger percentage of their total vote to the Democratic party's candidate than does the entire state, $\hat{P}_u$ will be too large.

Two simple alterations in the estimation method may improve our predictions. First, each of the state's 99 counties was assigned to one of 10 classes, or strata. Each stratum represents a cross-classification of historical voting record and current population (Table 1).

Table 1. County stratification

			Historical % G.O.P.*			
			Under 45%	45-50%	50-55%	Above 55%
County population	Under	25,000	4	19	35	18
	25,000 -	80,000	5	5	4	2
	Above	80,000	4	3	0	0

*   Average percent (of the electorate in each county) voting Republican in the previous three gubernatorial elections.

Thus stratum 1 consists of those 4 counties with "under 45% GOP" and "under 25,000" population, stratum 2 those with "45-50% GOP" and "under 25,000", etc. until stratum 10 contains those 3 counties with "45-50% GOP" and "more than 80,000" population. Note that there were no counties with more than 80,000 population and more than 50% GOP votes, so there are only 10 strata instead of 12.

Exercise

1.   What does this tell you in general about the results of the election in cities in the past?

The second alteration in the estimation method was to use the previous experience which had suggested that counties within a stratum would have similar voting patterns. Therefore, it was reasonable to use the vote information (at a given time t) from all of the precincts in a stratum to obtain appropriate estimates within each stratum. By assigning weights to the strata and combining the estimates from all of the 10 strata, one can obtain an overall estimate which may well be superior to $\hat{P}_u$. This alternative estimator may reduce the bias (because of the disproportionate representation of segments of the electorate) by properly weighting the data obtained from each of the strata.

EXAMPLE 1

Consider a simplified situation with only 2 strata and 7 counties. Suppose the information we have early on

election night is as shown in Table 2. Find the value
of the estimator, $\hat{P}_u$, of the percentage of the total
votes cast for the Democratic candidate. Can we improve on this estimate?

Table 2. Hypothetical data from 7 Iowa Counties

Stratum	County	n_c Number of precincts reporting	d_c Democratic votes cast	w_c Total votes cast
1	Linn	1	200	325
	Polk	2	300	425
2	Boone	1	150	250
	Cerro Gordo	1	200	400
	Jasper	1	300	600
	Kossuth	1	250	525
	Marion	1	100	225

Solution to Example 1. From the data given in Table 2,
the total Democratic vote (for all 7 counties) is 200 +
300 + 150 + ... + 100 = 1,500; and the total number of
votes cast is 325 + 425 + 250 + ... + 225 = 2,750.
Thus, $\hat{P}_u$ = (1500/2750) = 0.5455 = 54.55%.

   To determine an alternative estimate we first combine the data for the counties in each stratum.
(Recall that each stratum is composed of counties expected to have similar voting patterns.) Thus, in
Table 3, we summarize the data from Table 2 for the 2
strata. Here, for stratum c, n_c denotes the number of
precincts reporting; d_c denotes the number of Democratic votes reported; and w_c denotes the total number of
votes reported. To illustrate, we see from Table 2
that n_1 = 1 + 2 = 3, d_1 = 200 + 300 = 500, and
w_1 = 325 + 425 = 750.

Table 3. Hypothetical data from 2 Iowa strata

Stratum	Number of counties	Number of precincts	n_c	d_c	w_c
1	2	20	3	500	750
2	5	80	5	1,000	2,000

In the first stratum only 3 out of 20 precincts have reported with 500 Democratic votes out of a total of 750 cast while in the second stratum 5 out of 80 precincts have reported with 1,000 Democratic out of a total of 2,000. For our alternative estimate we want to take into account the fact that there are 80 precincts in Stratum 2 (which we assume are going to be much like the 5 already reported; i.e., about 50% Democratic) and only 20 precincts in Stratum 1 which seem to be going 2/3 Democratic.

We seem to be getting about $\frac{500}{3}$ or 167 Democratic votes per precinct in stratum 1 and so the 20 precincts should give us about $20(\frac{500}{3})$ Democratic votes. In stratum 2 we are getting $\frac{1,000}{5}$ Democratic votes per precinct and so we estimate $80(\frac{1,000}{5})$ for the whole stratum. This gives an estimate of 19333 Democratic votes. Similarly, for the total votes cast we estimate $20(\frac{750}{3})$ + $80(\frac{2,000}{5})$ or 37,000. The percentage of Democratic votes as estimated this way is then

$$\frac{19333}{37000} \quad \text{or} \quad .523 = 52.3\%$$

compared with the 54.55% we got before. If we extend this to all 10 strata in the whole state we get the following formula for computing $\hat{P}_s$ ("s" for "stratified").

Formally the stratified ratio estimator $\hat{P}_s$ is defined by

$$\hat{P}_s = \frac{\sum\limits_{c=1}^{10} N_c \bar{d}_c}{\sum\limits_{c=1}^{10} N_c \bar{w}_c} \, ,$$

where

N_c denotes the total number of precincts in the cth stratum,

d_c denotes the number of Democratic votes reported (at time t) in the cth stratum,

w_c denotes the total number of votes reported (at time t) in the cth stratum,

n_c denotes the number of precincts reporting (at time t) in the cth stratum

$$\bar{d}_c = \frac{d_c}{n_c} \qquad \text{and} \qquad \bar{w}_c = \frac{w_c}{n_c} \, .$$

Thus $\bar{d}_c$ is an estimator of the average number of Democratic votes per precinct for stratum c, and $N_c \bar{d}_c$ is an estimator of the total number of Democratic votes to be cast in stratum c. Finally,

$$\sum_{c=1}^{10} N_c \bar{d}_c$$

is an estimator of the total number of Democratic votes cast in all strata (i.e., throughout Iowa). Similarly,

$$\sum_{c=1}^{10} N_c \bar{w}_c$$

is an estimator of the total number of votes cast in Iowa.

Table 4. Data from the 1966 Iowa gubernatorial election: Harold Hughes (Democrat) against William Murray (Republican)

Stratum	GOP %	Size (in 000's)	Number of counties	Number of precincts N_c	n_c	d_c	w_c
1	0-45	0-25	4	76	10	793	1,48
2	45-50	0-25	19	409	45	3,439	6,25
3	50-55	0-25	35	710	87	8,445	16,49
4	above 55	0-25	18	347	12	965	2,52
5	0-45	25-80	5	167	8	711	1,14
6	45-50	25-80	5	149	7	613	1,27
7	50-55	25-80	4	107	13	1,494	2,99
8	above 55	25-80	2	61	11	2,114	3,38
9	0-45	above 80	4	284	17	3,454	5,00
10	45-50	above 80	3	166	16	2,192	5,03

Although this formula looks rather formidable, it is just an exercise in arithmetic if you follow the example for 2 strata worked out above.

Exercises for Example 1

Use the data of Table 4 to determine the uncorrected and stratified estimates for the Democratic vote percentages in the various cases. (Exercises 1, 2, 3, 4, 5 should be spread among the class, and their results combined for Exercise 6.)

1.   In strata 1 and 2.

2.   In strata 3 and 4.

3.   In strata 5 and 6.

4.   In strata 7 and 8.

5.   In strata 9 and 10.

6.   In the whole state.

7.   How would the uncorrected and stratified estimates compare if you considered any 1 stratum alone?

EXAMPLE 2

Table 5 displays the values of the estimators computed from the vote returns election night in 1964. On-the-air (WOI-TV, Ames, Iowa) predictions of final vote percentages were presented at 8:42 P.M., 9:05 P.M., 9:30 P.M., and 10:03 P.M. The first publicized estimate projected the landslide victory for incumbent Governor Harold Hughes. With 2.46% of the 2,476 precincts reporting, the stratified ratio estimator, $\hat{P}_s$, projected 0.6755 of the vote going for the incumbent against the final percentage of 0.6852. Note that the uncorrected percentage, $\hat{P}_u$, at the time of this report was 0.7219. The variations of each estimator ($\hat{P}_u$, $\hat{P}_s$) can be seen from the several projections presented in Table 5.

Table 5. 1964 gubernatorial election: percentage Democratic

% of precincts in	Time	$\hat{P}_u$ Uncorrected vote percentage	$\hat{P}_s$ Stratified ratio estimator	Final percentage
2.100		70.448	67.128	68.516
2.464	8:42 P.M.	72.186	67.547	68.516
8.320	9:05	68.564	66.938	68.516
13.732		67.959	65.583	68.516
15.913	9:30	68.682	66.608	68.516
19.103		69.046	67.001	68.516
28.998	10:03	69.505	67.921	68.516
37.197		69.669	68.036	68.516
42.932		70.209	68.739	68.516
56.018		69.756	68.677	68.516
83.683		68.847	68.578	68.516

From the state of Iowa official canvass:

Total Democratic	794,610
Total Republican	365,131
Total	1,159,741

The 1962 Iowa gubernatorial and senatorial elections, and the 1964 presidential election (restricted to Iowa) have also been studied. The results for these elections parallel those reported here. Further details are available in the two articles cited in the references.

Exercises for Example 2

1. Using the data in Table 5, plot the eleven values of $\hat{P}_u$ against the corresponding eleven values of "percent precincts in".

2. On the same piece of graph paper as you used in Exercise 1, plot the values of $\hat{P}_s$ against the corresponding values of "percent precincts in".

3. Using your results from Exercises 1 and 2 and the final vote percentage, what conclusions would you draw about the merits of using $\hat{P}_s$ rather than $\hat{P}_u$?

References

[1]   Robert W. Clyde, William J. Hemmerle, and T. A. Bancroft, "An application of 'post stratification' technique in local TV election predictions," Public Opinion Quarterly, 27 (1963): 467.

[2]   J. H. Sedransk and Robert W. Clyde, "A generalized technique for predicting election results from early returns," Public Opinion Quarterly, 30 (1966): 416.

The Last Revolutionary Soldier

A. ROSS ECKLER

INTRODUCTION

Can statistical analysis help to resolve a historical problem, particularly when the date are flawed and incomplete? This set considers one such problem and shows how to obtain an answer which, although uncertain and unprovable, has a reasonably high probability of being correct. The problem is the date of death of the last surviving soldier of the American Revolution. Some records give it as 1903, but this example shows that most probably the correct date is 1875, when Henry Gross died. How can we establish this, and how high is the probability that this date is correct?

The purpose of this example is threefold: (1) to show how statistical reasoning may be useful in a field (historical research) not ordinarily amenable to it; (2) to show how the method of analysis must be adapted to the availability and the quality of the data (often incomplete and flawed by errors) that are available; (3) to show that statistical data analysis is an iterative process (one must frequently carry out a sequence of analyses before the truth is finally found). Often it is unnecessary and unwise to use elaborate procedures of analysis when the data upon which the analysis is based are of doubtful accuracy and completeness--a common failing of historical information, which was not collected with the needs of a future statistician in mind--though sometimes elaborate methods are required

The author is at Bell Telephone Laboratories, Inc., Holmdel, New Jersey.

to make crude data yield any message. To illustrate these problems, consider what is apparently a simple problem--who was the last surviving American soldier of the Revolutionary War, and when did he die?

The problem appears simple because many people will recall the large amount of publicity generated by the last few survivors of the Civil War who died in the 1950's. However, the last survivors of the Revolutionary War received much less publicity, and, in fact, it is not entirely certain who held this honor. Two booklets on the subject were published at the close of the Civil War. Elias B. Hilliard interviewed six pensioned survivors in 1864, writing up the results in The Last Men of the Revolution (Hartford, Connecticut, 1865; reprinted by Barre Publishers, Barre, Massachusetts, 1968). G. W. Tomlinson wrote a small pamphlet entitled The Patriots of the Revolution of '76 (Boston, 1864), briefly describing 24 pensioners (12 from the North, 12 from the South) thought to be still alive in 1863. However, neither booklet lists Daniel Frederick Bakeman, who at the time of his death on April 5, 1869, was the last veteran of the Revolution receiving a pension from the U. S. Government.

But was he, in fact, the last survivor? Military records during the Revolution were so ill-kept that no one is certain how many men actually served in that conflict; estimates range from 184,000 to 250,000. Pensions were granted to those veterans who could offer satisfactory proof of service (receipts of pay, affidavits from contemporaries, and so forth) many years after the war; for example, those who served in the state militia were not in general eligible for pensions until Congress passed a law in 1832. Many of the survivors were undoubtedly unable to furnish satisfactory proof; others, perhaps, decided that the pensions were not worth the red tape involved in applying; and some may have been living in remote areas and were unaware that they were eligible for a pension.

It is possible that a search of local records, such as county histories or newspaper obituaries, might reveal who died after April 5, 1869. What is a plausible death date for the last survivor?

To examine this question, one needs a comprehensive list of Revolutionary soldiers and their dates of death. The only such list generally available is the DAR Patriot Index, published in 1967 and containing the names of about 100,000 Revolutionary soldiers having descendants who later joined the Daughters of the American Revolution. Although this book cannot be guaranteed to list the last survivor (who, after all, might not have had a descendant interested in the DAR), it can be used to construct a probabilistic model to estimate when the last survivor did die.

The following list gives the number of soldiers in the DAR Patriot Index dying in each year after 1860 (those men and women having patriotic service or civil service, as distinct from military service, are omitted).

1860	41	1870	2	1880	1	1890		1900	
1861	22	1871		1881		1891	1	1901	
1862	17	1872	2	1882		1892	1	1902	
1863	17	1873		1883	1	1893	1	1903	1
1864	11	1874		1884		1894	1		
1865	10	1875	2	1885		1895			
1866	6	1876		1886		1896			
1867	5	1877		1887		1897	2		
1868	1	1878		1888		1898	1		
1869	2	1879		1889		1899	1		

EXAMPLE 1

One should never consider analyzing data without examining them beforehand for sensibleness and consistency. Is there anything strange about the information in this table? What can be done about it?

Solution. The year 1903 occurred 120 years after the final year of the Revolutionary War; even assuming that the last soldier was a youthful drummer boy, he would have reached the age of 135 years. Despite claims of extreme longevity from places such as the Georgian Republic of the USSR (where reliable vital records, incidentally, are lacking), most demographers believe that ages of 135 or more are extremely unlikely to have ever occurred, and that the maximum span of human life is between 110 and 115 years. (For example, the Guinness Book of World Records claims that 113 is the longest well-authenticated life span.) It is far more likely that the DAR made errors in the compilation of the table; most of their work was done by DAR members or by clerical help, not by statisticians or other professionals trained in data collection and tabulation. Perhaps the quickest way to detect and eliminate such errors is to look at the ages of the soldiers at death (also published in the DAR Patriot Index), removing from the table all soldiers aged 120 or more years. (The choice of 120 is a somewhat arbitrary one; readers who believe that 120-year-old survivors are possible should pick 125 or 130 years as the cut-off age.)

If one goes back to the DAR Patriot Index and applies this rule, it is soon clear that 10 of the 13 soldiers dying in 1875 or later can be removed from the table. One of the others is listed as having been born in 1842 and can also be removed.

Fortunately, further cleaning of the data was possible. When these inconsistencies were pointed out to the compilers of the DAR Patriot Index, they checked all post-1870 deaths against the original membership cards and found numerous transcription errors, not just those that resulted in ages of 130 years or more. For example, the soldier who died in 1880 was 118 years old, not a completely impossible situation; however, it turned out that 1880 was a misreading of 1850. These corrections were included in the First Supplement to the DAR Patriot Index, published in 1969:

1860	41	1864	11	1868	2	1872	1
1861	23	1865	10	1869	2	1873	
1862	18	1866	6	1870	1	1874	
1863	17	1867	5	1871		1875	1

The moral of this story is that mistakes need not be perpetuated forever in a set of data; sometimes, one can reissue data in better form. The pattern of data in this table is much more reasonable. However, four soldiers in the DAR Patriot Index apparently outlived Bakeman: Henry Gross (d.1875), Jesse Harris (d.1872), Henry Smith (d.October 12, 1870), and Michael Burger (d.August 5, 1869).

EXAMPLE 2

Is Henry Gross a plausible last survivor?

Solution. In order to shed light on this question, it is useful to construct an idealized mathematical model of soldier mortality. Actuarial studies tell us that if we pick at random a person who has reached (say) his 100th birthday, he has a certain probability of living for one more year; in general, the older a person is, the smaller this probability becomes. We now make the following three assumptions in order to obtain a model of soldier deaths that is amenable to mathematical analysis:

1. Each soldier living at year x has the same probability $p(x)$ of surviving that year; furthermore, $p(x) > p(x+1) > p(x+2) > \ldots$.

2. Soldier deaths are independent from one year to the next; that is, the probability that a soldier will survive two years, given that he is living at year x, is the product of the probabilities $p(x)p(x+1)$.

3. Soldier deaths are independent of each other; that is, the fact that soldier A survives the year does not change the probability that soldier B will also survive the year.

It is clear that such a model can only approximately represent reality. For example, the value of p(x) depends upon other characteristics besides being alive at year x (some soldiers were in better health than others); furthermore, a wide-spread epidemic (such as the influenza epidemic of 1918) might result in non-independent soldier deaths in a given year. Nevertheless, because it is impossible to know the p(x) associated with each Revolutionary soldier, one must postulate a simplified model containing a single p(x) which is (in some cases) typical of the soldiers.

How does one obtain the probabilities p(x)? Numbers approximating these probabilities can be obtained from the list above. For example, 97 of the 138 soldiers alive on January 1, 1860, were also alive on January 1, 1861; therefore, the fraction of soldiers surviving the year 1860 was 97/138, or 0.70 to two decimal places. Similarly, 74 of the 97 soldiers alive on January 1, 1861, were also alive on January 1, 1862; therefore, the fraction of soldiers surviving the year 1861 was 74/97, or 0.76. The table below summarizes these fractions:

Year	Fraction of soldiers surviving the year
1860	0.72
1861	0.76
1862	0.76
1863	0.70
1864	0.72
1865	0.64
1866	0.67
1867	0.58
1868	0.71
1869	0.60

It is not hard to see that these fractions surviving are estimates--more or less accurate approximations--of the underlying probabilities of survival p(x). By analogy, if one tosses a coin 20 times and sees 11 heads and 9 tails, one estimates that the probability of heads is 11/20, or 0.55, even though the coin may be constructed in such a way that the true probability of heads is 0.50.

Notice that the fraction of soldiers surviving sometimes increases from one year to the next (for example, 0.58 of the soldiers survived 1867, but 0.71 survived 1868), whereas the sequence p(x), p(x+1), ... which these fractions are estimating has been assumed to be strictly decreasing. It is desirable to obtain

modified estimates of p(x), p(x+l), ... which also have
this property, and to extend the estimates to the years
1870, 1871, ... where the data are too sparse to
reliably estimate p(x).

Exercises

1.   Plot the fractions of soldiers surviving each year
      on a graph, with the years 1860, 1861, ..., 1869
      equally spaced along the horizontal axis; label
      additional years through 1875. (Note that it is
      only necessary to provide for a small part of the
      vertical axis--say, fractions surviving between
      0.50 and 0.75.) Notice that the points form a
      ragged pattern with a downward drift to the right.
      Using a ruler, draw a straight line through the
      region in which the points are clustered in such a
      way that the typical vertical distances from the
      points to the line are as small as possible (it may
      be helpful to think of the points as tiny weights
      imbedded on a sheet of cardboard, which is then
      balanced on a knife-edge). Read off the fractions
      surviving (the modified estimates of p(x), p(x+l),
      ...) on the vertical axis at the intersections of
      your straight line with the vertical lines repre-
      senting the years, 1860, 1861, ..., 1875.

      Statisticians have numerous methods for fitting a
line to data by using objective rules instead of sub-
jective judgment (as was done in the exercise). However, the incompleteness of the data and the fact that
our mathematical model is only a rough approximation to
reality suggests that we would be wasting our time
making a careful fit of a line to the data. In fact,
it even seems appropriate to simplify subsequent calculations by selecting a line which gives modified estimates of the survival probabilities containing only two
decimal places. This is done in the second column of
the table on the opposite page.

2.   Plot these points on the graph and connect them
      with a straight line. Is this line close to the
      one you drew before?

EXAMPLE 3.

How can we get the cumulative probability of survival
for each year?

Solution. These estimated probabilities can be combined, using the second assumption of the mathematical
model, to obtain the cumulative probability that a soldier alive on January 1, 1860, will survive 2, 3, 4,
etc. years. His probability of surviving several years

Year	Estimated probability of surviving the year	Cumulative probability of surviving all years
1860	0.72	0.72
1861	0.71	0.51
1862	0.70	0.36
1863	0.69	0.246
1864	0.68	0.168
1865	0.67	0.112
1866	0.66	0.074
1867	0.65	0.048
1868	0.64	0.031
1869	0.63	0.0194
1870	0.62	0.0121
1871	0.61	0.0074
1872	0.60	0.0044
1873	0.59	0.00260
1874	0.58	0.00151
1875	0.57	0.00086

is the product of his survival probabilities for each individual year; thus, he has a probability of (0.72)(0.71) = 0.51 of living to January 1, 1862, and a probability of (0.72)(0.71)(0.70) = 0.36 of living to January 1, 1863. These cumulative probabilities are tabulated in the third column of the above table.

If we know the number, n, of Revolutionary soldiers alive on January 1, 1860, we can use this table of cumulative probabilities to obtain a plausible death date for the final survivor. If one multiplies each of the cumulative probabilities by n, one has an estimate of the number of survivors at the end of that year. It is clear that the number of estimated survivors decreases year by year and may be made arbitrarily small.

It may seem plausible to say that the last survivor has died when the number of expected survivors has been reduced to zero. However, with a little reflection one sees that the number, a product of positive numbers, will never reach zero no matter how many years are considered. Similarly, it is not reasonable to say that the last survivor has died when this number has been reduced to one, for he is still alive then. When the number has been reduced to 0.5, an acceptable compromise is reached--the single survivor has a probability of 1/2 of surviving the time-interval corresponding to the reduction of the number of estimated survivors from 1.0 to 0.5.

EXAMPLE 4

The only problem remaining is the estimation of n, the number of Revolutionary soldiers alive on January 1,

1860. Unfortunately, this is difficult to obtain. We propose two different estimates of n, one of which seems likely to be too high, the other too low.

Solution. According to the DAR Patriot Index (and its First Supplement), 52 of the 138 soldiers alive January 1, 1860, were granted pensions. Furthermore, it is easy to determine that 17 of the 24 pensioners listed by Tomlinson are also listed in the DAR Patriot Index. If one assumes, using Tomlinson's figures, that the 52 DAR pensioners represent 17/24 of the total pensioner population on January 1, 1860, then there must be $(7/24)(52) = 15$ pensioners not listed in the DAR Patriot Index who should be added to the 138 soldiers already known. Thus, one can get n equal to 153. Obviously, this is a lower bound, for it ignores those Revolutionary soldiers who neither received a pension nor had a DAR descendant.

The DAR Patriot Index lists approximately 100,000 soldiers, whereas the World Almanac estimates that 184,000 to 250,000 men participated in this conflict. The midpoint of these extremes is approximately 215,000; let us inflate the DAR survivors by a factor of 2.15 to obtain $138(2.15) = 297$ survivors on January 1, 1860. This estimate is almost certainly too high; one can argue that the soldiers in the DAR Patriot Index, being family men, may have had better care and a more settled existence, and consequently a greater life expectancy.

If one multiplies the successive cumulative probabilities of surviving all years (given in the table above) by n = 297, one eventually discovers that there were $(297)(0.00151) = 0.45$ survivors as of January 1, 1875. If one multiplies the successive cumulative probabilities by n = 153, there were $(153)(0.00260) = 0.40$ survivors as of January 1, 1874. In other words, the last survivor appears to have died late in 1874 or late in 1873. This compares very well with the 1875 death rate of Henry Gross; it seems quite plausible that he was the last survivor.

Exercise

3.   The reasonableness of this mathematical model of soldier mortality can be tested by estimating the date of the last survivor of the Civil War. Unfortunately, no one has bothered to collect information about Civil War soldiers comparable to the DAR Patriot Index. However, on May 30, 1949, Life magazine published the names and pictures of 68 Civil War survivors. Assuming that no soldiers were overlooked in the Life article and assuming the same mortality behavior for Revolutionary and

Civil War soldiers, construct a table of cumulative probabilities of survival for 1949, 1950, 1951, ..., 1960. [Hint: The year 1949, 84 years after the end of the Civil War, corresponds to the year 1867, 84 years after the end of the Revolutionary War.] Using this table of cumulative probabilities, estimate the death date of the last Civil War survivor.

Answer

Year starting May 30	Estimated probability of surviving the year	Cumulative probability of surviving all years
1949	0.65	0.65
1950	0.64	0.416
1951	0.63	0.262
1952	0.62	0.162
1953	0.61	0.099
1954	0.60	0.059
1955	0.59	0.0350
1956	0.58	0.0204
1957	0.57	0.0116
1958	0.56	0.0065
1959	0.55	0.00357
1960	0.54	0.00192

If one multiplies 68 by 0.0065, one obtains 0.44; therefore, the last soldier died shortly before May 30, 1959 (ten years after the Life article).

How does this estimate compare with the actual death date of the last Civil War soldier? Walter J. Williams, who died on December 30, 1959, was widely publicized as the last survivor. However, his claim to have served in the Civil War was later proved to be false--he was not present in the 1850 Federal Census and was only five years old at the time of the 1860 Census. At present, the last Civil War soldier is believed to have been John Salling, who died on March 17, 1959, at about the age of 112. This agrees remarkably well with the prediction from the table above. This strengthens our belief that the mathematical model is a reasonable one to use.

We now return to the Revolutionary soldier problem. In assessing the quality of a mathematical model supposed to represent reality, it is important to learn whether or not the answer given by the model is sensitive to small changes in the numerical values used in the model.

How sensitive is this typical date of the last survivor to the straight-line fit on the probability of survived data? To get an idea, assume that the probability of survival has a constant value of 0.7 (probably too high), or 0.6 (probably too low). In the former case, the typical date of the last survivor is delayed about three years, and in the latter case it is advanced about one and one-half years.

There is an even more important source of variation which we have so far ignored. Each soldier's date of death is a random event, and it is exceedingly unlikely that exactly 0.72 of the soldiers will survive the first year, (0.72)(0.71) the first two years, and so on. What we are more interested in knowing is the following: if each soldier has a probability of dying each year according to the above formula, how much variation in the date of death of the last survivor can one plausibly expect? If a hypothetical population of n soldiers were repeatedly exposed to these probabilities of death, between what two limits would most of the final death dates fall? It is only when we have information of this sort that we can truly decide the plausibility of Henry Gross's claim, or know what range of dates to search in old newspapers or other contemporary records.

Class project

The following exercise should be carried out independently by each student in the class, and the answers obtained by different students compared at the end in order to get some notion of the variability of the death date of the last survivor. What we are going to do is simulate by means of random numbers the actual pattern of year-by-year deaths of 153 Revolutionary soldiers. Enter a table of random numbers at a random page; suppose that the digits read 32018475591105... . View these digits as successive pairs: 32, 01, 84, 75, 59, ... Assign a one-digit pair to each of the 153 soldiers; if the pair is 01, 02, ..., 72, the soldier is assumed to survive the year 1860, but if the pair is 73, 74, ..., 99, 00, the soldier is assumed to have died in the year 1860. (Each student should use a different set of random numbers.) Count up the number of soldiers surviving the year 1860 and assign a fresh digit-pair to each of these soldiers. If the pair is 01, 02, ..., 71, the soldier is assumed to survive the year 1861, but if the pair is 72, 73, ..., 99, 00, the soldier is assumed to have died in the year 1861. (Note that the dividing-line has been changed to reflect the reduced probability of surviving the year 1861.) Count up the number of soldiers surviving the

year 1861, and assign a fresh digit-pair to each of
these soldiers. Continue in this fashion until a year
is reached in which no soldiers survive; this is the
year of the last survivor. How close to 1873, the
theoretical estimate derived earlier, did you come?
[Note: Students who have a digital computer available
may wish to program it to carry out this simulation.
It should be relatively easy to obtain several hundred
death years for the final survivor; these can then be
plotted in a histogram to get some idea of the variability of the simulation.]

Index

Index

73